Math Survival Guide

Tips and Tricks for Science Students

Math Survival Guide

Tips and Tricks for Science Students

JEFFREY R. APPLING
Clemson University

JEAN C. RICHARDSON
Parkview High School

SECOND EDITION

WILEY

JOHN WILEY & SONS, INC

ACQUISITIONS EDITOR	Deborah Brennan
MARKETING MANAGER	Robert Smith
PRODUCTION EDITOR	Barbara Russiello
ILLUSTRATIONS	Jeffrey Appling
COVER DESIGNER	Madelyn Lesure
COVER PHOTO	Renee Lynn/Photo Researchers, Inc.

This book is printed on acid-free paper.

To order books or for customer service, please call 1-800-CALL WILEY (225-5945).

ISBN 0-471-27054-7

10 9 8 7 6 5 4 3 2 1

PREFACE

What type of calculator should you buy before you start your first year of college science? How do you estimate a logarithm without using a calculator? What study technique has raised exam scores by more than 30 points? Who is Oscar, and why the heck was he trying to catch salmon? In this book we answer these questions and provide other ideas and strategies that might help students in their first college science courses.

Students sometimes need to brush up math skills in preparation for a first year science course. This book is meant as a review to help those students as well as students presently in science courses who need a quick reference to basic math techniques. It is not a text, and it does not cover every aspect of useful mathematics. Rather, it is a "survival guide," where basic ideas have been gathered together for discussion. Sometimes that discussion is brief and efficient, and sometimes the discussion has more depth in order to get across important concepts. Either way this book should help you if you feel a little shaky about your math skills. In addition, we have included study tips and insights that will help you in all of your science courses. By exploring these ideas you will develop the necessary skills to become a successful science student.

If you are preparing to take your first college level science course you should read this book from the beginning. After the term progresses you can use this book as a resource to remind you of forgotten methods and to learn tips and tricks for inventing your own problem-solving shortcuts. You probably know much of the math contained in this book. What we have tried to write is a summary packed with hints that will bring your math skills up to speed efficiently.

This book starts with the basics of numbers and their manipulations, including ratios, powers, roots, logarithms, scientific notation, and significant figures. With numbers under control, the reader can advance into the territory of equations, units, and graphing. A short review of trigonometry and geometry precedes chapters aimed at improving problem solving and study skills. The reader is introduced to helpful methods (estimation techniques, for example) that also enhance understanding. These later chapters target ways to study more efficiently and with better retention. We have included these chapters to make this survival guide more than a math summary and to provide ways for you to develop good study habits. We compiled these tips by observing students—the good ones. Finally, we have included exercises and answers in the appendices. Do these problems to strengthen your skills and try out new tricks that you learn in the chapters.

In this second edition we have added references to solving problems using a TI-83 Plus graphing calculator. If mathematics is a struggle for you, you might find the TI-83 Plus a very user-friendly calculator, especially given the keystrokes and graph screen images that we have provided in the book.

We wrote this book for students, and we are genuinely interested in what students think about it. Let us know how you feel about *Math Survival Guide.* If you have any pet tips or tricks that we can add, send them to us via electronic mail. We look forward to hearing your thoughts.

Jeffrey R. Appling   Jean Richardson

japplin@clemson.edu   jean_richardson@gwinnett.k12.ga.us

Clemson, SC   Lilburn, GA

January 2003

CONTENTS

CHAPTER 1

Your Calculator: Friend or Foe?

It sure seemed like a good idea at the time. To tackle the imposing mountain of science, you bought a state-of-the-art calculating machine. A miniature computer. To be your helper, your companion, your weapon against the sinister forces of math. It had buttons. *Lots* of buttons. You named it Hal.

But here you are in the middle of an exam and Hal has deserted you. You can't remember how to take a cube root, even though you know there is a menu somewhere for doing just that. You try to graph an equation, but where did the axes go? You chide yourself for letting your roommate talk you into downloading games instead of learning how to use your new calculator. Your roommate turned the axes off to play a game and you don't know how to turn them back on! Ten minutes have been wasted going through different menus. You try to work some of the problems by hand and hope for the best.

After the test you approach your instructor and report that your calculator caused you problems. As brief discussion ensues, you hope for some special consideration, but your instructor merely probes you about what happened. You grudgingly reveal that the graphing calculator works, but that you don't know very much about your new friend. Your instructor picks up the calculator, and hands it back after a few keystrokes. That cube root is under the Math menu and the Format key will turn on the axes. Two simple keys! Your heart sinks.

The preceding scenario need not occur. Having problems with your calculator happens for only one reason—you don't really know how it works. It is now easy to afford a powerful handheld calculating machine and a reasonable sum brings you mathematical wizardry in a box. But the average student does not have the time to plow through the 100-page owner's manual, so most people wing it with their new toys. The math that you will use in most first year science courses is relatively simple. With the exception of trigonometric functions, roots, and logs, you will be able to do most calculations by hand. But the sophisticated calculator goes beyond this—and it can help you if you just become its master.

The solution is easy: No matter how tempting they are, with their graphic screens and alphanumeric memories, do not buy a sophisticated calculator UNLESS you are willing to really learn how the darn thing works. If your classes don't require a graphing calculator, buy a calculator more suited to your real, current needs. There are quite a few models that will do the job for less than $20. Check out the nearest office warehouse for a nice selection. When you do take a class that requires a graphing calculator, spend time learning how to use it—long before your first quiz.

So what functions does your "nerd-o-lator" need for a science course? They all have the big four: addition, subtraction, multiplication, and division. You will need exponential notation (so the freebie from the bank won't really cut it) and at least one memory location. Memories are sometimes shown as **STORE/RECALL** keys. A key to allow changing the sign (from plus to minus and vice versa like **±**, **(–)**, or **CHS**) comes on most models. An **INVERSE** key (often **INV**) will allow you to do the inverse operation of certain functions. The **INVERSE** function is essential when doing trigonometric calculations like sine and cosine. A key labeled x^{-1} or $1/x$ will allow you to take the reciprocal of numbers and fractions. For example, the easy way to get the decimal value of one-fifth is to enter the number **5** followed by the **1/x** key (two keystrokes) to yield **0.2**, which is the same result you get from the four keystrokes: **[1 ÷ 5 =]**. Throughout this book we will use bold brackets (**[ ]**) to show a sequence of calculator keystrokes on a nongraphing calculator, and bold braces (**{ }**) for graphing calculators.

For simple trigonometry, the function keys **SIN, COS,** and **TAN** (coupled with their inverses) are sufficient. A calculator equipped with these functions will also have the capability to do calculations in either degrees or radians (portions of **pi**) and will have a way to switch between the two. There will also be a key to quickly bring up the value of **pi** (3.1415926535897932384 . . .).

You will need functions based on exponents (powers) and their inverses. For instance, to easily **SQUARE** a number, the x^2 key is handy. The **SQUARE ROOT** key, $\sqrt{\ }$, will do just the opposite. For powers other than two, the key labeled x^y allows any combination. You enter x first, followed by the power to which x is raised. For example, suppose you need the fourth power of ten: 10^4. You would enter [**10** x^y **4** =], which would yield the desired answer: **10,000**. Taking roots other than square roots is just as straightforward. The cube root ($\frac{1}{3}$ root) of one million would be [**1000000** $x^{1/y}$ **3** =], which yields the result **100**. The $x^{1/y}$ key allows for fractional exponents.

The powers of ten are common in scientific calculations, so a 10^x key is useful. To determine exponents for numbers in base 10, you use a function called the logarithm. A **LOG** key designates this function. Other useful operations in science are done in "base e," where e is a specific irrational number. Power functions with e are done with an e^x key, and the opposite function, the natural logarithm, uses the **LN** key. We will discuss powers and logarithms in more detail in Chapters 4 and 5.

So for less than $20, you can have a small nongraphing calculator that will get you through science class with a minimal investment. You should know, however, that many students will have a graphing calculator, and it is very likely that your professor will have one as well. It can be much harder to input data into the small nongraphing calculator as you are essentially typing things in backwards. For instance, if you want to take the square root of 81, you type 81 in first, then press the square root key. The display shows only the last entry, while a graphing calculator display shows multiple lines so you can easily see your expressions—and they don't disappear after you evaluate them.

Graphing calculators are the way of the future. Not only are they changing the way we calculate, they are changing the way teachers teach, too. Because it is easy to see graphs, charts, and tables, teachers can extend their ideas further and expect more from their students. A graphing calculator can do virtually any math required for your math and science courses and will probably be appropriate for use all through college and more! It is easy to input data into graphing calculators, and you can type it in exactly as you see it on the board or in your textbook. You will make fewer mistakes if you can see what you are typing. Be sure to research all calculators before making that big purchase, then familiarize yourself with the calculator you buy by using it to do your homework prior to any exams. The more money you spend, the more the calculator will do—but it may be less user-friendly as a result.

We are going to use the TI-83 Plus in our discussions in this book. There are several reasons why the TI-83 Plus is our recommendation for a graphing calculator. It is extremely user-friendly, and many high school and college students are using them in their math and science courses (for more than just games!). Texas Instruments has a Web site that supports all their calculators and offers tutorials for getting started with a graphing calculator.

We will explore some of the manipulations available on your calculator, including some shortcuts. The aim of this book is to review the math necessary to do well in first year science courses. Math is the language used for solving problems. The next chapter will introduce some letters of this language—numbers.

Chapter 1 Summary

- A nongraphing calculator should at least have these keys:

  $+ \; - \; \times \; \div \; =$

  EXP (or **EE**) for scientific notation

  STO (or **M**) for memory

  $\pm$ **INV** $1/x$ (or x^{-1})

  π **SIN COS TAN**

  $x^2 \; \sqrt{} \; x^y \; x^{1/y} \; 10^x \; \textbf{LOG} \; e^x \; \textbf{LN}$

- If you do buy a powerful calculator, *learn how to use it!*
- Consider a TI-83 Plus graphing calculator. It is fairly easy to use and may be all you ever need for your math and science courses. Texas Instruments has a Web site for support, http://education.ti.com

Practice Exercises Chapter 1 ———————————————

Try these problems to exercise your calculator skills. Calculate an answer for the quantity shown.

1. $\dfrac{3.53 + 6.20}{3.53 - 6.20}$

2. $\dfrac{1}{4(3.5) - 6.0}$

3. $(4.12)^3 - 2.33 \log(6.2 - 2.6)$

4. $5 + \sqrt{(5)^2 - 4(1)(6)}$

5. $\dfrac{4.7 \times 10^{-3} + 2.3 \times 10^{-3}}{6.5 \times 10^{-7}}$

6. Evaluate $x^2 - 5x + 6$ for $x = 3$

7. Evaluate $\dfrac{(2x)^2}{(0.10 - x)(0.10 + x)}$ for $x = 5.0 \times 10^{-2}$

8. $\sin(\frac{\pi}{2}) + \cos(\frac{\pi}{2})$

9. $\tan^{-1}(1.00)$ in degrees

10. Show a complete graph for $y = \dfrac{x - 2}{x + 3}$

11. Show a complete graph for $y = 4x^3 - 74x^2 + 300x$

12. $\sqrt[4]{(1296)^3}$

13. Determine $e^{-\left(\frac{E}{RT}\right)}$ for $E = 5000$, $R = 8.31$, and $T = 298$

14. Find $f'(2)$ for $f(x) = 2x^2 + 1$

15. A child jumps off a 12-meter platform into a pool. The height in meters above the water is modeled by $f(x) = -4.9t^2 + 12$, where t is the time in seconds since the child jumped. How long will it take for the child to be 3 meters above the water?

CHAPTER 2

Numbers

Now that you have a good, dependable calculator, you'll need some numbers to put into it. Even though success in science courses will depend on your grasp of the concepts, you will often have to show your abilities through the use of numbers. So it will pay for you to master them. In this chapter you will get reacquainted with some old friends.

By the age of three you could probably count to 10. Those **counting numbers** were all positive (unless you were a very strange child), but some time later you learned the concept of zero. The **whole numbers** include zero. It was probably much later that you added the negative numbers to complete the number line:

Counting Numbers: $\{ 1, 2, 3, 4, 5, \ldots +\infty \}$

Whole Numbers: $\{ 0, 1, 2, 3, 4, 5, \ldots +\infty \}$

Integers: $\{ -\infty \ldots -5, -4, -3, -2, -1, 0, 1, 2, 3, 4, 5, \ldots +\infty \}$

These numbers are **integers**. The symbol ∞ represents **infinity** (a concept that is difficult to fathom whether you are 3 or 30).

At some point you learned about money. There were quarters, nickels, dimes, and pennies—all parts of a dollar. With a money system based on 100 pennies in a dollar, you soon learned about the **decimal** system. A price tag of $2.98 meant 298 pennies. Using the decimal point, you could easily show fractions of dollars. Now there were other "numbers" between the integers.

The next set of numbers includes fractions. A **fraction** represents a value between two whole numbers. Numbers that include the integers and fractions between them are called the set of **rational numbers**. Rational numbers cannot be listed like the integers because there are so many. Here are two definitions of a rational number:

I. Rational Number: **A number is rational if it can be put into the form $\frac{a}{b}$, where a and b are integers and $b \neq 0$.**

- 7 is rational; it can be written as $\frac{7}{1}$, or as $\frac{21}{3}$ (or many other ways).

- 0.4 is rational; it can be written as $\frac{4}{10}$ or $\frac{2}{5}$.

- $2\frac{7}{8}$ is rational; it can be written as $\frac{23}{8}$.

II. Rational Number: **A number is rational if it can be written as a decimal that either repeats or terminates (stops).**

- $\frac{1}{2}$ is rational; it can be written as a decimal (0.5) that terminates.

- $\frac{2}{3}$ is rational; its decimal equivalent (0.6666…) repeats.

Fractions are written as one integer divided by another. To convert a fraction to a decimal number, simply divide the upper number (**numerator**) by the lower number (**denominator**). A quarter is, of course, $\frac{1}{4}$. Dividing **1** by **4** will give **0.25**. Fractions greater than 1 can be represented as either a straight ratio (more on ratios in Chapter 3) or a **compound fraction**. For example, $\frac{23}{8}$ is equal to the compound fraction $2\frac{7}{8}$. Learn to convert compound fractions quickly; to convert $2\frac{7}{8}$ the numerator is **7 + (8 × 2) = 23**. The new denominator is identical to the old one in the compound fraction. You can also think of $2\frac{7}{8}$ as $2 + \frac{7}{8}$, or $\frac{16}{8} + \frac{7}{8} = \frac{23}{8}$, since 2 is equal to $\frac{16}{8}$.

Numbers in science are often decimal but are not restricted to two places after the decimal. For instance, if you measured a mass to the nearest milligram, it might have the value 2.137 grams. The significance of how many positions are occupied in a number (like the difference between 2.137 and 2.13700) will be covered in detail in Chapter 7. In Chapter 6 we will discuss **scientific notation**, the most common and useful way of presenting very big or very small numbers.

Most numbers that you will encounter will have specific values. However, there are two important exceptions. The first you are probably already familiar with, **pi** (π). Pi has a value close to 3.14159265358979, but it is an **irrational number** so the decimal equivalent has an infinite number of places. You might encounter π in calculations using angles in geometry and trigonometry (covered in Chapter 11). The other common irrational number is *e*. Base *e* is important in scientific calculations; it is the base of the **natural logarithm** (more on logarithms in Chapter 5). You can always get a value for *e* (2.71828...) since your calculator has an [*e*] key on it. Here is a correct, although slightly cheesy, definition of an irrational number:

Irrational Number: An irrational number is a number that is not rational.

- $\sqrt{7}$, π, and *e* are irrational; their decimal equivalents do not repeat or terminate.

- $\sqrt{7}$, π, and *e* cannot be expressed in the form $\dfrac{a}{b}$ (*a* and *b* are both integers).

```
√(7)
          2.645751311
π
          3.141592654
e
          2.718281828
```

The rational numbers and the irrational numbers make up the numbers that we use mainly in day-to-day life and in scientific and mathematical calculations. These numbers may be all you ever need in your lifetime. This broad set of numbers is called the Real Number System:

Real Numbers: The combination of the rational numbers and the irrational numbers make up the Real Number System.

- 12 is real because 12 is also rational.

- $\sqrt{11}$ is real because $\sqrt{11}$ is also irrational.

The numbers discussed so far, with the detail provided in subsequent chapters, cover common calculations that you will encounter in your first few years of science coursework. Eventually you will need to manipulate **complex numbers**, and they are mentioned here for completeness. A complex number is a combination of two parts. The first part is a **real number**, like the integers, decimals, and fractions already discussed. The second part consists of an **imaginary number** (a **real coefficient** times i, where $i = \sqrt{-1}$). The complex number is thus written as two parts like **2 + 3i**, or more generally, **a + bi**. Here is a definition of complex numbers:

Complex Numbers: The imaginary numbers combined with the real numbers, in the form a + bi, where a is the real part and bi is the imaginary part, make up the Complex Number System.

- 2 + 3i is complex; 2 is the real part and 3i is the imaginary part.
- 12 is complex; it can be written as 12 + 0i, with 12 being the real part and 0i the imaginary part.
- $\sqrt{-12}$ is complex; it is the same as $\sqrt{-1 \cdot 4 \cdot 3} = 2i\sqrt{3}$. This is a pure imaginary number because the real part is 0.

Figure 2.1 shows the relationship between the numbers discussed.

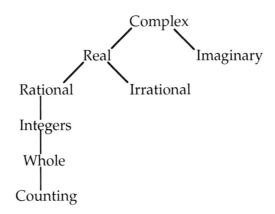

Figure 2.1 Relationships between number systems.

- −12 is an integer, a rational number, a real number, and a complex number.
- $\sqrt{14}$ is an irrational number, a real number, and a complex number.
- $\frac{7}{4}$ is a rational number, a real number, and a complex number.
- 6 is a counting number, a whole number, an integer, a rational number, a real number, and a complex number.

Chapter 2 Summary

- **Integers** include the **whole numbers** and the negative **counting numbers**.
- **Rational** numbers are integers and the numbers between the integers that are commonly expressed as decimals or fractions.
- The **irrational numbers**, **π** and *e*, are used in special types of calculations.
- Numbers commonly used in scientific calculations are **real numbers,** including **rational** and **irrational numbers**.
- Most calculations are conducted with **real numbers**, although **complex numbers** are important in advanced scientific work.

Practice Exercises Chapter 2 ————————————————

1. Convert these compound fractions to simple fractions and decimal numbers:

   a. $5\frac{5}{9}$ c. $3\frac{5}{11}$

   b. $2\frac{23}{32}$ d. $-4\frac{2}{7}$

2. Reduce these fractions to yield simple fractions with the smallest numerators and denominators:

   a. $\dfrac{36}{84}$ c. $\dfrac{92}{116}$

   b. $\dfrac{26}{39}$ d. $\dfrac{91}{259}$

3. Convert these simple fractions to compound fractions and decimal numbers:

   a. $\dfrac{13}{8}$ c. $-\dfrac{56}{26}$

   b. $\dfrac{305}{137}$ d. $\dfrac{256}{16}$

4. What are the decimal equivalents for these common simple fractions?

   a. $\frac{1}{8}$, $\frac{3}{8}$, $\frac{5}{8}$, $\frac{7}{8}$

   b. $\frac{1}{3}$, $\frac{2}{3}$, $\frac{4}{3}$, $\frac{5}{3}$

5. Convert these decimal numbers to simple fractions and compound fractions (see exercise 3 for help):

   a. 13.375   c. $-4.3\bar{3}$

   b. 7.625   d. $3.6\bar{6}$

6. Use your calculator to determine values for the irrational numbers π and e.

7. Classify each number in as many ways as possible, using counting, whole, integer, rational, irrational, real, imaginary, and complex numbers:

   a. -22 c. -6.2 e. $0.\overline{85}$ g. $6 + 2i$

   b. $\sqrt{36}$ d. $\sqrt{15}$ f. $\frac{3}{2}$ h. $0.70700700070000\ldots$

8. Put these real numbers in order from smallest to largest:

   $\frac{28}{9}, e, \pi, \sqrt{11}, 2\frac{5}{7}, \sqrt{4}$

9. Put the following real numbers in order from smallest to largest:

   $-3, -\sqrt{10}, -\frac{16}{5}, -\sqrt[5]{e^6}, -\pi, -\frac{7}{2}$

10. Express each rational number in the form $\dfrac{a}{b}$, where a and b are integers and $b \neq 0$. Use three different expressions for each.

    a. 2.7

    b. $1\frac{3}{5}$

CHAPTER 3

Ratios

Think about the first day of one of your classes—history, for example. It is natural to look around and learn about your classmates. Let's say you count 60 women and 40 men (including yourself). When asked about your class by your nosey roommate, you might respond "There are 60 women and 40 men." Or you might express the numbers as a **ratio**, a term often used to relate the amounts of things.

To retain the original numbers of students in your class, you might tell your roommate "The male-to-female ratio is 40 to 60." You could write this **40:60**, using a colon (**:**) as the symbol to indicate a ratio. It would be more common, however, for you to mentally reduce the ratio of 40:60 by dividing by a common factor. You could relate the same information with the ratio **4:6** (dividing by 10), or more simply as **2:3** (dividing by 2 again, or dividing the original ratio by 20). Your roommate would then know that there were three women students for every two men.

A ratio is a comparison of two numbers.

The ratio of a to b can be written in three ways:

a to b $\qquad\qquad$ $a{:}b$ $\qquad\qquad$ $\dfrac{a}{b}$

- $\frac{40}{60}$ is the male to female ratio , while $\frac{60}{40}$ is the female to male ratio.

- 60:100 is the female to total student ratio.

- $\frac{3}{5}$ is the female to student ratio in simplest form $\left(\frac{60}{100} \div \frac{20}{20} = \frac{3}{5}\right)$.

Ratios are useful in science. They can give relations between items in a quantitative way. A common use for ratios is as "conversion factors" (unit conversions will be discussed in detail in Chapter 9). For example, you know that there are 12 inches in 1 foot. The ratio of inches:feet is therefore 12:1. You can also represent the same thing as

$$\frac{12 \text{ inches}}{1 \text{ foot}}$$

where the ratio symbol has been replaced by the familiar division line. You might read this as "there are 12 inches per foot." By including the units of inches and feet, the method is more compact. You don't need to preface the ratio of 12:1 with the explanation "inches:feet." We will use this format in the chapter on units and their conversion.

Comparing values for measured amounts often involves the manipulation of a **proportion**.

A proportion is an equation formed by equating two ratios.

If two ratios are equal, their **cross products** are equal:

$$\text{If} \quad \frac{a}{b} = \frac{c}{d}, \text{ then } ad = bc.$$

Using the class ratio example, 2:3 = 40:60 represents one proportion that can be stated about your history class. While this is a proper proportion, you will see it most often written in the format:

$$\frac{2}{3} = \frac{40}{60}$$

These are equal ratios because each cross product is 120. In other words, the product $2 \times 60 = 3 \times 40$.

Proportions are useful when you know a ratio but would like to scale up or down. You might ask the question "If I make $8.00 per hour, how much money will I make in a 40-hour week?" This can be expressed as a proportion with an unknown quantity that you want to know. Notice in the proportion that each ratio is written in terms of dollars ($) to hours:

$$\frac{\$8}{1\,\text{hr}} = \frac{x}{40\,\text{hr}}$$

This equation can be satisfied when the unknown quantity is equal to $320, since $8:1 = $320:40. Of course, you can arrive at the answer by cross multiplying to determine that $x = 8 \times 40 = \$320$. When you compute how much you will make in a given time, you are solving a proportion by cross multiplying. We will work more with equations in Chapter 8 and use proportions to help convert units in Chapter 9.

One very important issue that involves equations and proportions is the idea of **proportionality**. Consider the density of objects, whether they are "heavy" or "light." The density, D, is determined by the object mass (m) and its volume (V):

$$\text{Density} = \frac{\text{mass}}{\text{Volume}}$$

$$D = \frac{m}{V}$$

For a given volume (size), as the object mass increases, the density increases—the object gets "heavier." This type of relation (D increases when m

increases) is a *direct* proportionality. The density and mass are said to be **directly proportional**. The relationship between density and volume is just the opposite. For a given mass, as the volume increases, the density *decreases*. You would say that a coin with mass 2.2 kilogram (1 pound) is heavy, whereas a 2.2 kilogram piano would seem extremely light. Density and volume are **inversely proportional**, since D decreases as V increases. This behavior happens because D is in the numerator of the left side of the equation and V is in the denominator of the right side. As V gets larger (coin $\rightarrow$ piano), $\frac{m}{V}$ gets smaller (you are dividing by a larger number) and hence D decreases.

Let's go back to your first day of history class. Perhaps your nosey roommate doesn't phrase the question as "What is the male-to-female ratio?" An equivalent question could be "What fraction of your history class is women?" This is a request for the same information but in a different form. Recall that the ratio was $\frac{60 \text{ women}}{40 \text{ men}}$. The **fraction** of women is equal to the number of women related to *the sum* of all the students in the class (presumably only in the categories "men" and "women"—no "other"). You must relate the number of women to the addition of women plus men:

$$\frac{60 \text{ women}}{60 \text{ women } + \text{ 40 men}} = \frac{60 \text{ women}}{100 \text{ students}} = 0.60$$

However, it is unlikely that your roommate would understand your response as "zero point six."

How do you communicate that 60 out of 100 students are women? Simply use a **percentage.** You will naturally convert 60 women per 100 total students to **60%**.

A percent is a unique ratio that compares a number to 100. Percent means "per one hundred."

Therefore, the number 60% is equal to $\frac{60}{100}$, which is also 0.60. To find the percentage of something, change the fraction to a decimal and multiply by 100%. So the fraction of women in your history class is 0.60, and the percentage of women is 60%. Therefore the fraction of men is $1 - 0.60 = 0.40$, and the percentage of men is $100\% - 60\% = 40\%$.

Any number represented as a percentage shows the "parts per one hundred." Most totals won't be 100, of course, so you must first find the decimal and then convert to percentage. For example, if your biology class has an enrollment of 195, and 4 students graduated from your high school, you and your friends make up about 2% of the class:

$$\tfrac{4}{195} = 0.0205, \, 0.0205 \times 100\% = 2.05\%$$

To find the percent of a number, you can also write a proportion using this relationship:

$$\frac{\text{part}}{\text{whole}} = \frac{\text{part}}{\text{whole}}$$

In this relationship, the percent is the numerator, the part, and 100 is the denominator, the whole. Using the biology class example from above,

$$\frac{4}{195} = \frac{x}{100}$$

Cross multiply: $4 \times 100 = (195)(x)$, or $400 = 195x$. To solve for x, divide both sides of this equation by 195:

$$\frac{400}{195} = \frac{195x}{195}$$

This yields $x = 2.05$, 400 divided by 195. Therefore, the proportion becomes:

$$\frac{4}{195} = \frac{2.05}{100} = 2.05\%$$

Remember that any percentage can be converted to its decimal equivalent for use in calculations. To change a percent to a decimal you simply move the decimal point two places to the left. Conversely, to change a decimal to a percent, move the decimal two places to the right. Don't forget to include the percent sign. For example, if 13% of your calculus class made an "A" on the first exam, and there are 54 students in the class, you can determine the number of students that made an "A": 13% of 54 = 0.13 × 54 = 7.02. So seven students made an "A." Note that "13% of" means to multiply as shown previously.

An important consequence of using percentages is that the sum of all percentages must equal 100%. Think of 100% as 100/100, which is equal to 1. All parts of a whole must sum to equal 1. You can use this to get yourself out of a jam sometimes. It may appear that a problem is missing information, when in fact you can determine the missing number by remembering that the sum of all parts of a whole must equal 1 (or all percentages of a whole must sum to 100%). Consider the following problem:

Analysis of a hydrocarbon containing carbon, hydrogen, and oxygen yields 59.99% carbon and 4.48% hydrogen. What is the empirical formula for this compound?

In order to begin this problem, you need the percentage of oxygen, which is not given. Don't panic; you can easily solve for the missing number:

59.99% C + 4.48% H + **?% O** = 100%

(All percentages must sum to 100%.)

% O = 100% − (59.99% C + 4.48% H) = 100% − 64.47% = **35.53% O**

(This compound is aspirin, for those of you playing along at home.)

Say your fruit fly colony of 1272 flies has 14 mutants. You determine the fraction of mutants as 14/1272 = 0.011 and convert to percentage by multiplying the answer by the number 1 (remember, you can always multiply anything by the number 1 without changing its value). In this case you might choose the number 1 to be 100%: $0.011 \times 100\% = 1.1\%$ mutants. Alternatively, you can divide the fraction, move the decimal two places to the right, and add a % sign.

What about another expression of the number 1? How about in "parts per thousand," also known as **ppt**. Just as 100% is equal to 1, so 1000 ppt (1000/1000) is also equal to 1. This allows you to represent your mutants in another way: $0.011 \times 1000 \text{ ppt} = 11 \text{ ppt}$ mutants. In other words, for every group of 1000 flies, 11 will be mutants. Both forms are acceptable—and equal to one another. You can use "parts per million," **ppm**, in a similar fashion. In this format, $0.011 \times 1,000,000 \text{ ppm} = 11,000 \text{ ppm}$. This number is bulky, so you might choose to use percentage or ppt for a streamlined result. Parts per million and parts per billion, **ppb**, are typically used for small concentrations of species in water (toxins for example).

Chapter 3 Summary

- **Ratios** relate amounts of two things, *a* and *b*, in the format *a:b* or $\frac{a}{b}$.
- Equal ratios have equal **cross products**.
- A **proportion** is an equation where ratios are equated. To solve a proportion, cross multiply and solve for the unknown quantity.
- A **fraction** relates one thing to the whole. Common ways of expressing fractions include **percentage (%)**, **parts per thousand (ppt)**, **parts per million (ppm)**, and **parts per billion (ppb)**.
- The sum of all fractions of a whole must equal **1**. In addition, the sum of all percentages of a whole must equal 100%.
- To change a decimal to a percent, move the decimal two places to the right (include the % sign!). To change a percent to a decimal, move the decimal point two places to the left (leave off the % sign).

Practice Exercises Chapter 3 ———————————————————

1. Simplify these ratios, and convert them to fractions:

    a. 20:120

    b. 13:52

    c. 2.4:17.2

    d. 2:12,000

2. Solve each proportion:

    a. $\dfrac{x}{12} = \dfrac{15}{180}$

    b. $\dfrac{3}{15} = \dfrac{x}{6}$

3. Change each decimal to a percent:

    a. 0.76

    b. 1.21

    c. 0.0046

    d. 0.084

4. Change each percent to a decimal and into a fraction in lowest terms:

    a. 85%

    b. 8.2%

    c. $32\frac{1}{2}\%$

    d. 3%

5. Change each ratio to a percent:

    a. $\dfrac{1}{8}$

    b. $\dfrac{2}{5}$

    c. $\dfrac{4}{7}$

    d. $\dfrac{16}{23}$

6. Use a proportion to find the percent for the given ratio:

    a. $\dfrac{4}{121}$

    b. $\dfrac{1}{6}$

    c. $\dfrac{7}{25}$

    d. $\dfrac{5}{9}$

7. Find each number:

    a. 16.1% of 180

    b. 75% of 2552

    c. $2\frac{1}{2}\%$ of 88

    d. 110% of 6721

8. Find each number by setting up a proportion and solving:

   a. 24% of 180   c. 2% of 112

   b. 85% of 12   d. 125% of 426

9. At Homecoming, there are 45,000 fans rooting for the home team and 5000 fans supporting the visiting team. What is the ratio of Home Fans:Visiting Fans? What percentage of the stands are filled with visiting fans?

10. Your ecology class counted a total of 74 reptiles and amphibians in a one-square-mile area near a pond. If 22% of the animals were reptiles, how many amphibians were counted?

11. Four out of five dentists agree that toothpaste is overpriced. (We made that up, can you tell?) At a convention of 632 dentists, how many dentists think toothpaste is overpriced? What percentage think that toothpaste is *not* overpriced?

12. Your new job pays $47,500 per year. You work 50 forty-hour weeks in a year. What is your salary by the week? By the hour?

13. A certain soap claims to be 99 and $\frac{44}{100}$% pure. In a ton of soap (2000 pounds), how much "impurities" are there?

14. When heated to 200 °C, a 120-inch metal rod lengthens by $\frac{3}{8}$ inch. By what percentage does the rod elongate? By what ppt?

15. A local reservoir is polluted with PCBs at a concentration of 20 ppm by volume. The reservoir holds 10 million gallons of water. What volume of PCBs is present?

CHAPTER 4

Powers and Roots

The world of science is riddled with shorthand notation. This is not done to confuse outsiders but rather to streamline discussions of complex ideas. The math that scientists use is often likewise streamlined to make manipulations of numbers and equations easier to fathom. Some mathematical descriptions fit this use of simplified notation. The next three chapters cover important examples: **powers (exponents)**, **roots**, **logarithms**, and **scientific notation**.

Use of **powers** arises naturally from the need to express a series of multiplications. For example, the number 64 results from multiplying the number 2 times itself six times:

$$2 \times 2 \times 2 \times 2 \times 2 \times 2 = 64$$

It is certainly easier to express this operation using the conventional notation 2^6. This shorthand style reads "two raised to the sixth power;" the number **2** is the **base** and the number **6** is the **exponent**. This expression instructs you to multiply 2 times itself six times. So one may write:

$$\mathbf{2^6 = 64}$$

which is more concise than the original equation. In general, the expression a^n is a power of a, where a is the base, n is the exponent, and $a^n = a \times a \times a \times a \ldots \times a$, n times.

The series of 2 raised to greater and greater values of the exponent is referred to as "powers of two." Two raised to the zero power, 2^0, is defined equal to **1**. In fact, *any* number (except 0) raised to the power zero is equal to **1**. The series thus becomes:

$$2^0 = 1, \, 2^1 = 2, \, 2^2 = 4, \, 2^3 = 8, \, 2^4 = 16, \, 2^5 = 32 \ldots$$

The powers of two are very important in the science of computing. Computers "think" in 0's and 1's—the only digits necessary in base 2.

A very important series, one that is used throughout science, is the powers of ten:

$$10^0 = 1, \, 10^1 = 10, \, 10^2 = 100, \, 10^3 = 1000, \, 10^4 = 10{,}000 \ldots$$

The milestones represented by each power of ten are called **decades**, or **orders of magnitude**. For instance, if two experiments yield data that differ by two orders of magnitude, then they are a factor of 100 off from one another. (Back to the drawing board!) Our decimal number system is based on these powers of ten: The digits in a number represent ones, tens, hundreds, thousands, and so on. You will learn in Chapter 6 about **scientific notation**, an important representation of numbers using powers of ten.

If an exponent can be equal to zero, can it be negative? Of course—negative whole number exponents yield numerical values less than **1**. The negative sign instructs you to divide the number into **1**:

$$a^{-n} = \frac{1}{a^n}$$

For example,

$$10^{-1} = \frac{1}{10^1} = \frac{1}{10}$$

$$4^{-5} = \frac{1}{4^5} \text{ and } \frac{1}{4^{-5}} = \frac{1}{\frac{1}{4^5}} = 4^5$$

$$\left(2^4\right)(-3)^{-5} = \frac{2^4}{(-3)^5}$$

Note that numbers with negative exponents are the reciprocals of their positive exponent counterparts, for example, $4^{-5} = \frac{1}{4^5}$.

As with most mathematical operations, **powers** have an inverse operation. The number 3 raised to the second power ("three squared") equals 9. The inverse operation is to take the number 9 and ask "the square of what number will equal 9?" To answer this question you need to take the **second root**, or the **square root**, of 9. You can represent this algebraically using the equation

$$x^2 = 9$$

You must solve for the value of x using the square root operation. The **radical symbol** $\sqrt{}$ represents taking the root of a number: $\sqrt{x^2} = \sqrt{9}$. For the second root (square root), the index 2 is not used: $\sqrt{9} = \sqrt[2]{9}$. Note that $(-3)^2 = 9$ as well, so $\sqrt{9}$ actually has two values, -3 and $+3$. Your calculator will report only the positive root.

Since exponents can have values larger (and smaller, as you will soon see) than 2, you must also be able to take roots larger than 2. And such is the case in the following examples. Referring to our example from the beginning of the chapter, the sixth root of 64 is equal to 2:

$$\sqrt[6]{64} = \sqrt[6]{2 \cdot 2 \cdot 2 \cdot 2 \cdot 2 \cdot 2} = 2$$

The sixth root of 64 is a number that when multiplied by itself 6 times produces the given number 64. We know that $2^6 = 64$, so the answer is 2. The cube root of a number, $\sqrt[3]{x}$, is a number that when multiplied by itself 3 times produces the given number x. So, $\sqrt[3]{64} = \sqrt[3]{4 \cdot 4 \cdot 4} = 4$, because $4^3 = 64$.

The nth root of a number, $\sqrt[n]{x}$, is a number that when multiplied by itself n times produces the given number x.

Thus, taking the roots of numbers is the inverse operation of raising numbers to exponents.

Can you have exponents that are not integers? Again the answer is yes. The simplest examples are exponents that are fractions. Fractional exponents represent roots—a number raised to a fraction actually indicates that a root should be determined. Consider 9 raised to the one-half power:

$$9^{\frac{1}{2}} = 3$$

The exponent $\frac{1}{2}$ instructs us to take the second root of 9, so $9^{\frac{1}{2}}$ is another way of writing $\sqrt{9}$. Fractional exponents equal the roots: $64^{\frac{1}{6}} = \sqrt[6]{64} = 2$. This will become clearer when multiplication and division of exponential numbers is discussed later in the chapter. The following general expressions hold:

$$a^{\frac{1}{n}} = \sqrt[n]{a}$$

$$a^{\frac{m}{n}} = \sqrt[n]{a^m} = \left(\sqrt[n]{a}\right)^m$$

Some examples include:

$$8^{\frac{1}{3}} = \sqrt[3]{8} = 2$$

$$27^{\frac{4}{3}} = \left(\sqrt[3]{27}\right)^4 = 3^4 = 81$$

$$4^{-\frac{3}{2}} = \left(\sqrt{4}\right)^{-3} = 2^{-3} = \frac{1}{2^3} = \frac{1}{8}$$

That exponents can be fractions implies they can have *any* decimal value. This is absolutely true and is easily checked with your calculator. For example try to solve **5.03⁻¹·²³**: {**5.03 ^ (–) 1.23 ENTER**}. Did you get 0.137? You could not do that example by hand, although calculating powers by hand can be done for whole number exponents. Calculating roots is much more difficult, so you will rely on your calculator for these operations (naturally, committing some perfect squares and cubes to memory occurs through exposure and practice—quick, what is the cube root of 8?). Your calculator makes taking roots and most powers easy. Here are some examples:

$9^{\frac{1}{9}} = 1.277$ {**9 ^ (1 ÷ 9) ENTER**}

$7^{-2} = 0.0204$ {**7 ^ (–) 2 ENTER**}

$e^{3} = 20.086$ {***e* ^ 3 ENTER**}

$2^{-\frac{1}{\pi}} = 0.802$ {**2 ^ ((–) 1 ÷ π) ENTER**}

Multiplication and division of numbers with exponents can often be done by hand, even though you will probably still rely on your calculator to confirm your answer. This is a waste of time since you are actually doing the calculation twice! Instead, learn to recognize the simple calculations that can be done quickly by hand (or more correctly, in your brain), so that you skip writing steps and solve problems faster. Let's first look at the mechanism for multiplying and dividing these numbers.

Multiplication of exponential numbers can be done quickly when the number being raised to the exponents (called the "base," remember?) is the same, for example **2³ × 2⁶** (here the base is **2**). The answer is obtained by *adding* the exponents (3 + 6 = 9) to obtain the new exponent:

$$(2 \times 2 \times 2) \times (2 \times 2 \times 2 \times 2 \times 2 \times 2) = 2^9$$

$$2^3 \times 2^6 = 2^{3+6} = 2^9$$

Thus the result is 512. How about when exponents are negative? Try this one: $10^{-3} \times 10^3$. The answer is 1. Add the exponents: $-3 + 3$ is zero, so 10^0 is 1. In general:

$$a^m \; a^n = a^{m+n}$$

For example,

$$3^{-4} \times 3^6 = 3^{-4+6} = 3^2 = 9$$

$$2^{-2} \times 2^{-3} = 2^{-2+(-3)} = 2^{-5} = \frac{1}{2^5} = \frac{1}{32}$$

Division of numbers with exponents is similar to multiplication. Instead of adding exponents, you *subtract* them. For instance, the correct exponent derived from $2^3 \div 2^2$ is 1:

$$\frac{2^3}{2^2} = 2^{(3-2)} = 2^1 = 2$$

In general,

$$\frac{a^m}{a^n} = a^{(m-n)}$$

You subtract the second exponent from the first. Actually, you may find it easier to convert a division operation into a multiplication operation. Remember that the negative exponents represent reciprocals, like $\frac{1}{5^2} = 5^{-2}$.

This means that any division operation can be converted to a multiplication operation by changing the sign on the exponent of the number in the denominator:

$$\frac{3^2}{3^{-4}} = 3^2 \times 3^4 = 3^{(2+4)} = 3^6$$

Note that for the second number you just change the sign (or multiply by –**1**, if you prefer) when you change the division to multiplication.

This trick might make a difference if you never can remember which exponent to subtract from which when doing a division operation.   By converting to a multiplication operation, you are left with the less complicated addition of exponents. Granted, you are using extra steps—and possibly time—to do the conversion. But if you have trouble organizing the manipulations in your mind, use the approach that gives you the right answer reliably.

Here are a few other properties of exponents that may come in handy:

$$\left(a^{m}\right)^{n} = a^{m \times n}$$

This means that if you have a power raised to a power, you multiply the exponents. For example, $(2^{3})^{4} = (2^{3})(2^{3})(2^{3})(2^{3}) = 2^{(3+3+3+3)} = 2^{12}$ or $2^{3 \times 4}$. Also,

$$(ab)^{n} = a^{n} \cdot b^{n} \text{ and } \left(\frac{a}{b}\right)^{n} = \frac{a^{n}}{b^{n}}$$

For example,

$$\left(7x^{2}\right)^{3} = 7^{3}x^{6}$$

$$\left(\frac{3x^{2}}{4y^{-3}}\right)^{2} = \frac{3^{2}x^{4}}{4^{2}y^{-6}} = \frac{9x^{4}y^{6}}{16}$$

Chapter 4 Summary

- Powers and roots are shorthand notation for complex multiplication (powers) and division (roots).
- Exponents can be any number—positive, negative, integer, or real.
- Negative exponents represent reciprocals:

$$10^{-5} = \frac{1}{10^5} \text{ and } \frac{1}{2^{-3}} = 2^3.$$

- Fractional exponents represent taking roots:

$$8^{\frac{1}{3}} = \sqrt[3]{8} \text{ and } 32^{0.25} = \sqrt[4]{32}.$$

- To multiply exponential numbers with the same base, add the exponents:

$$3^{2.5} \times 3^{0.5} = 3^3 \text{ and } 10^{-3} \times 10^{-5} = 10^{-8}.$$

When multiplying numbers with different bases, you must evaluate the exponential numbers separately before performing the multiplication.

- To divide exponential numbers, subtract the exponent of the divisor (the denominator) from the exponent of the dividend (the numerator):

$$5^3 \div 5^{16} = 5^{3-16} = 5^{-13} \text{ and } 10^3 \div 10^{-6} = 10^3 \times 10^6 = 10^9.$$

- The general behavior of exponents includes:

$$a^{-n} = \frac{1}{a^n} \qquad\qquad a^{\frac{m}{n}} = \sqrt[n]{a^m} = \left(\sqrt[n]{a}\right)^m$$

$$a^m \times a^n = a^{m+n} \qquad\qquad \frac{a^m}{a^n} = a^{(m-n)}$$

$$(a^m)^n = a^{m \times n} \qquad\qquad (ab)^n = a^n \cdot b^n$$

$$\left(\frac{a}{b}\right)^n = \frac{a^n}{b^n}$$

Practice Exercises Chapter 4 ————————————

1. Convert these numbers to exponential numbers:

    a. 100,000 (base 10) c. 0.0001 (base 10)

    b. 512 (base 2)   d. 625 (base 5)

2. Evaluate these numbers by hand, and verify the answers with your calculator:

    a. 7^4 d. 2^{-2}

    b. $(1.5)^3$ e. $\sqrt[3]{125}$

    c. 10^7 f. $(121)^{\frac{1}{2}}$

3. Determine these values with your calculator:

    a. $(2.2)^5$ d. $\sqrt[3]{\frac{\pi}{2}}$

    b. $(5.3)^{-2}$ e. $(4.21)^{-0.23}$

    c. $8^{-\frac{1}{3}}$ f. $e^{-\pi}$

4. Perform these calculations:

    a. $2^3 + 2^4$ c. $\sqrt{\pi^3 + \pi}$

    b. $10^3 - 10^2$ d. $\dfrac{10 - 10^2}{10 + 10^2}$

5. Multiply:

    a. $x^7 \cdot x^4$ c. $10^5 \cdot 10^{-\frac{1}{5}}$

    b. $2^{-\pi} \cdot 2^{-1}$ d. $2x^{-4}y \cdot 3x^7y^5$

6. Divide:

a. $\dfrac{2^6}{2^3}$

c. $\dfrac{x^7 y^4}{x^6 y^2}$

b. $\dfrac{10^{-2}}{10^4}$

d. $\dfrac{27 x^4 y^3}{3 x y^2}$

7. Simplify and express with positive exponents only:

a. 10^{-3}

c. $x^{-7} \cdot x^{-3}$

b. $\dfrac{2}{7 e^{-\pi}}$

d. $\dfrac{3^{-7}}{3^{-2}}$

8. Express each in radical form first (root, $\sqrt[n]{x}$), then simplify:

a. $8^{\frac{1}{3}}$

c. $64^{-\frac{2}{3}}$

b. $9^{\frac{3}{2}}$

d. $32^{\frac{3}{5}}$

9. Express each in exponential form and simplify if possible:

a. $\sqrt{81}$

c. $\sqrt[4]{16 x^4 y^6}$

b. $\sqrt[3]{3^7}$

d. $\sqrt[6]{7^{-3}}$

10. Simplify:

a. $\left(3^2\right)^4$

c. $\left(\dfrac{4x}{y}\right)^{-2}$

b. $\left(2 x^2 y\right)^3$

d. $\left(27^{\frac{1}{3}}\right)^4$

Logarithms

Logarithms seem to give students headaches. Maybe this chapter will be just the medicine you need. Logarithmic functions define many measurement scales in science, including pH (acidity), decibel (sound level), and Richter (earthquake intensity) scales. For example, an earthquake that registers 6.3 on the Richter scale has an amplitude ten times larger than an earthquake at 5.3. A difference of one unit in the scale reflects a tenfold change since the logarithmic scale compresses the orders of magnitude. Logarithms are useful in dealing with extremely large and extremely small numbers—particularly when you must deal with them together.

Suppose you had to develop a shorthand way of expressing numbers that could have the values 10^0, 10^1, 10^2, 10^3, 10^4, 10^5, or 10^6. An easy way to identify each number would be by the exponents 0, 1, 2, 3, 4, 5, or 6. You have just invented the logarithm. A **logarithm** is the exponent to which a base number (2, 10, e, etc.) is raised to yield a specific result; in other words, a logarithm is the inverse function of exponentiation (see Chapter 4).

To better understand the relationship between exponents and logarithms, we need to focus on the behavior of inverse functions. Consider the function $y = 2^x$. This is an **exponential function** in the form $y = a^x$, where a is a positive real number other than 1, and x is any real number. Graphs of exponential functions include the point (0,1) as shown:

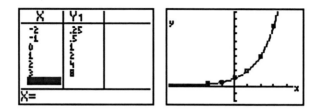

One way to show the inverse of $y = 2^x$ is to switch the x- and y- coordinates in the table of ordered pairs:

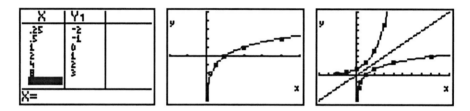

What is the equation for the inverse of $y = 2^x$? If you switch the variables, the inverse is $x = 2^y$. To solve this equation for y, a logarithm is needed—we can write an exponential function in an equivalent logarithmic form. Hence, $y = \log_2 x$ **is the same as** $x = 2^y$ and is the inverse of $y = 2^x$. Note that graphs of inverse functions reflect in the line $y = x$, as shown in the preceding calculator screen.

It might be useful to think of it this way: The equation $x = 2^y$ is the inverse of $y = 2^x$. Therefore, if we take the logarithm of $x = 2^y$, we must do it to both sides of the equation:

$$\log_2(x) = \log_2\left(2^y\right)$$

On the right side, the log function returns the exponent of the base 2,

$$\log_2(x) = y$$

Turning the resulting equation around, we have the relation $y = \log_2 x$.

In general,

$$y = \log_a x \text{ means } a^y = x$$

where a is a positive real number (called the base) other than 1. If there is no base, it is understood to be 10. In other words, $\log(100) = 2$ means $10^2 = 100$. The [LOG] key on your calculator means base 10 (**common log**) and the [LN] key means base e (**natural log**). Logarithms in base 10 (log) and base e (ln) are most often used in scientific applications.

- $\log_2(8) = 3$ because $2^3 = 8$

- $\log_{10}(10{,}000) = 4$ because $10^4 = 10{,}000$

- $3^4 = 81$ means $\log_3(81) = 4$

- The inverse of $y = 10^x$ is $x = 10^y$, or $y = \log_{10} x$

- $\ln(e^{-5}) = -5$

As you can see from these examples, a logarithm is just an exponent! The preceding examples are obvious but not encountered that often. Usually you will take the logarithm of a number not raised to a power. For instance, what is $\log(70)$? Remember that this means $10^x = 70$. In other words, you are looking for the exponent to which you raise 10 in order to get 70. You know that $10^1 = 10$ and $10^2 = 100$, so the answer must be a number between 1 and 2. Using your calculator, just type in {**LOG 70 ENTER**}. You will see that the output $\log(70) = 1.84509804$, and $10^{1.84509804}$ equals 70.

```
log(70)
           1.84509804
10^1.84509804
                   70
```

Because $y = \log(x)$ and $y = 10^x$ are inverses, these functions will cancel the effects of each other. To show this on your calculator, type in {**LOG 70**} then press {**2ⁿᵈ 10ˣ 2ⁿᵈ ANS**}. The {**ANS**} key recalls the last number in your calculator, which was 1.84509804. This keeps you from having to type the number in again and prevents rounding errors. Your display should look like this:

```
log(70)
          1.84509804
10^(Ans)
                 70
```

Because exponential functions and logarithmic functions are inverses, the following properties are true:

For $b > 0$ and $b \neq 0$,

$$\log_b b^x = x$$

$$b^{\log_b x} = x$$

For example,

- $10^{\log(3)} = 3$

- $\log_4\left(4^5\right) = 5$

- $\ln\left(e^2\right) = 2$

In your studies, you might encounter the **antilogarithm**, another term for the inverse of a logarithm. The shorthand notation is **antilog**. The inverse of $y = \log x$ is $y = 10^x$, so $y = \text{antilog}(x)$ means the same thing as $y = 10^x$. Now wasn't that simple? Generally,

$$\text{antilog}(x) = 10^x$$

For example,

- antilog(2) means $10^2 = 100$

- antilog(–4) means $10^{-4} = 0.0001$

- antilog(–1.33) means $10^{-1.33} = 0.0468$

Sometimes you may need to convert between logarithms in base 10 and natural logarithms. In order to do this, we use the **change of base property** so that we can change a log with any base to any other base:

For any positive real numbers $a \neq 1$, $b \neq 1$, and $x > 0$:

$$\log_b x = \frac{\log_a x}{\log_a b}$$

For example,

- $\log_3 x = \dfrac{\log x}{\log 3}$

- $\ln x = \dfrac{\log(x)}{\log e} = 2.303 \log(x)$

The change of base property is also useful if you need to graph $y = \log_b x$ since most calculators only have keys for base 10 and base e logarithms.

There are some properties of logarithms that may be useful when it is necessary to rearrange an equation that contains a logarithm. These properties are written with base b, which means they can be used for common logs (base 10), natural log (base e), or any other base:

For $m > 0$, $n > 0$, b > 0, r any real number and b $\neq$ 1,

Product Property: $\log_b(mn) = \log_b m + \log_b n$

Quotient Property: $\log_b \dfrac{m}{n} = \log_b m - \log_b n$

Power Property: $\log_b m^r = r \log_b m$

For example,

- $\log(6) + \log(2) = \log(6 \times 2) = \log(12)$

- $\log(6) - \log(2) = \log\left(\frac{6}{2}\right) = \log(3)$

- $\ln\left(6^2\right) = 2\ln(6)$

- $\log_3(54) - \log_3(2) = \log_3\left(\frac{54}{2}\right) = \log_3(27) = 3$ because $3^3 = 27$

Keep in mind that the following relationships are **NOT** true:

- $\log(a + b) \neq \log(a) + \log(b)$

- $\dfrac{\log(a)}{\log(b)} \neq \log\left(\dfrac{a}{b}\right)$

Chapter 5 Summary

- Logarithms are used to find unknown exponents.
- $y = \log_a x$ means $a^y = x$.
- The inverse of a logarithmic function is an exponential function.
- Antilogarithms are the inverse of logarithms; $\text{antilog}(x) = 10^x$.
- On your calculator, use the [LOG] key for common logs and the [LN] key for natural logs. (The inverse functions, 10^x and e^x, are usually located above the [LOG] and [LN] keys.)
- Common logs are related to natural logs by the equation $\ln(x) = 2.303\log(x)$.

Practice Exercises Chapter 5 ──────────────────────────

1. Find an equation for the inverse function. Graph both to check that the graphs reflect in the line $y = x$.

   a. $y = \frac{1}{3}x + 1$ b. $y = 5^x$

2. Write each logarithm in exponential form:

   a. $\log_8(64) = 2$ c. $\ln(1) = 0$

   b. $\log(1000) = 3$ d. $\log\left(\dfrac{1}{10}\right) = -1$

3. Write each in exponential form and solve without a calculator:

   a. $\ln\left(e^{-3}\right)$ c. $\log_4(64)$

   b. $\log\left(10^6\right)$ d. $\log(0.01)$

4. Use a calculator to evaluate the logarithm. Check using the inverse operation.

   a. $\log(25)$ c. $\ln(3)$

   b. $\log(0.23)$ d. $\ln(100)$

5. Use the properties of logarithms to express each as a single logarithm. Evaluate without a calculator.

   a. $\log(100) + \log(10)$ d. $\log_4(32) + \log_4(2)$

   b. $\log_5(75) - \log_5(3)$ e. $\ln\left(e^4\right) - \ln\left(e^3\right)$

   c. $3\log(5) + 3\log(2)$ f. $5\ln e - 2\ln e$

6. Use the change of base property to change to a common logarithm:

   a. $\ln(5)$ b. $\log_3(12)$

7. Use the change of base property to change to a natural logarithm:

    a. $\log(62)$ b. $\log_6(17)$

8. Change the function to a common log using the change of base property. Graph the function on your calculator: $y = \log_4 x$.

9. Change the function to a natural log using the change of base property. Graph the function on your calculator: $y = \log_4 x$. Compare your result to that for exercise 8.

10. Simplify using the inverse properties:

    a. $e^{\ln(2)}$ c. $e^{2\ln(3)}$

    b. $10^{\log(6)}$ d. $10^{2\log(4)}$

CHAPTER 6

Scientific Notation

Our universe is a mixture of the very big and the very small; you will learn this fact early in your study of science. Consider, for example, the differences in the mass of the Earth and the mass of one molecule of oxygen. The best current values of these quantities are:

mass of Earth = 5,980,000,000,000,000,000,000,000 kg

mass of oxygen molecule = 0.0000000000000000000000000531 kg

(kg are kilograms, a unit of mass). It is not convenient to write out these long numbers! If it were part of your job to deal with very large or very small numbers you would soon throw up your hands in disgust—unless you turned to a shorthand notation for help. You can relax. This shorthand exists, in the form of **scientific notation** (also called **exponential notation**). Scientific notation expresses a number as a product of a number between 1 and 10, and a power of 10.

Let's think of a number that's a little more exciting than the mass of the Earth. What if you graduate from college and are rewarded with a gift from your rich uncle? How about a BMW 540i Sport, list price $53,300? There are other ways to represent this number that will introduce the use of exponential notation. First, move the decimal point so that the original number is now a number between 1 and 10 (there should be one digit to the left of the decimal). Use the symbol ˘ to represent where the decimal point should be:

$$5 \smallsmile 3300 \text{ is a number between 1 and 10}$$

Next, count the decimal places from where you put the new decimal point to where the decimal point was originally (no decimal point means it is understood to be at the end). This will be the exponent of 10, and it will be positive because we counted to the right. Therefore, scientific notation for 53,300 is

$$5.3300 \times 10^4$$

There, you just did scientific notation. In this case it doesn't seem warranted—you are familiar with dollar amounts in the $50,000 range (or at least you *intend* to become accustomed to them after graduation!). Now let's put the mass of Earth into scientific notation using the method described in the preceding example.

mass of Earth = 5,980,000,000,000,000,000,000,000 kg

$5 \smallsmile 980,000,000,000,000,000,000,000$ is a number between 1 and 10

Next, count from the new decimal point to where the original decimal point was (at the end of the number). You should count 24 places to the right; therefore, scientific notation for the mass of the Earth is

$$5.98 \times 10^{24} \text{ kg,}$$

which is much easier to use than the longhand version of the number.

Now try the same method with the mass of the oxygen molecule.

mass of oxygen = 0.00000000000000000000000000531 kg

0.0000000000000000000000000005 ˇ 31 (5.31 is a number between 1 and 10)

Count from the new decimal point to the original decimal point. You should count 26 places to the left, which means the exponent of 10 will be –26. Therefore, the scientific notation will be:

$$5.31 \times 10^{-26} \text{ kg}$$

The large, negative exponent should jump out at you—this is a *very* small number. Remember that multiplying by 10^{-26} means to multiply by $\frac{1}{10^{26}}$, or to divide by 10^{26}.

Your graphing calculator will do scientific notation in a very straightforward way. Press the Mode key and select scientific mode as shown:

```
Normal SCI Eng
Float 0123456789
Radian Degree
Func Par Pol Seq
Connected Dot
Sequential Simul
Real a+bi re^θi
Full Horiz G-T
```

Press {2nd **QUIT**} to go to a clear screen and just type in the number and press {**Enter**}. Let's type in the mass of the Earth. Your screen will look like this:

```
5980000000000000
000000000
          5.98E24
```

As you are probably aware, 5.98E24 means 5.98×10^{24}.

So there you have it; big numbers have large positive exponents and small numbers have large negative exponents. What other types of manipulations are made with these new shorthand numbers? You will most likely rely on your calculator for standard operations with exponential numbers, but you should understand how to make these maneuvers yourself—for two reasons. First, you can do a lot of calculating *without* your calculator to save time (see also discussion in Chapter 13, Making Estimates). Also, there may be an occasion when your calculator cannot handle the desired calculation. (Horror of Horrors! Well, after all, it *is* only a machine) For example, pull out your nerdulator and try this one:

$$\left(4.87 \times 10^{56}\right) \div \left(2.23 \times 10^{-45}\right)$$

{(4.87 EE 56) ÷(2.23 EE (−) 45) ENTER}

So what happened? If your calculator is like most, you got the big **ERR: OVERFLOW**. That's the symbol the calculator uses when the result of your request is beyond the memory abilities of the circuitry. Your little handheld friend isn't perfect after all.

You must prepare for this unfortunate limitation of your calculator—preferably *before* a class quiz. To solve the preceding equation with your calculator you need one of the rules for manipulation of numbers expressed in scientific notation. Let's look again at the calculation:

$$\left(4.87 \times 10^{56}\right) \div \left(2.23 \times 10^{-45}\right)$$

Rewrite it this way and solve:

$$\frac{4.87 \times 10^{56}}{2.23 \times 10^{-45}} = \frac{4.87}{2.23} \times 10^{(56-(-45))} = 2.18 \times 10^{101}$$

So your answer should be 2.18×10^{101}, not **ERR**.

Actually, this is not the most common type of scientific notation mistake that you may make. Most mistakes happen when addition or subtraction is attempted. The rule to keep in mind is that to add or subtract numbers represented in scientific notation, you must have *both numbers* represented using the *same power of ten*. This method is not a problem for your calculator, yet you should master it for future estimation problems and for determining the proper number of significant figures—the topic of the next chapter. You wouldn't add dollars to pennies without first converting the dollars to pennies ($2 + 33 cents = 200 cents + 33 cents = 233 cents) or vice versa ($2 + 33 cents = $2 + $0.33 = $2.33).

Consider the addition of 2.13×10^5 and 7.55×10^3. Choose the number with the smallest exponent and rewrite the other number using that power of ten: $2.13 \times 10^5 = 213. \times 10^3$ (note that the decimal was moved two places to the right). Now you can add the two numbers:

$$
\begin{array}{r}
213. \times 10^3 \\
+\ 7.55 \times 10^3 \\
\hline
220.55 \times 10^3
\end{array}
$$

This may make more sense if you remember some of your algebra skills. Let's factor out the 10^3:

$$213. \times 10^3 + 7.55 \times 10^3 = 10^3(213. + 7.55) = 10^3(220.55)$$

Now you can write the sum as 220.55×10^3, or change it to 2.2055×10^5 (decimal moved 2 places to the left). But wait—the original numbers only had 3 digits each, and this result has 5. Is this right? Tune in to the next chapter to see how to represent this answer properly.

Chapter 6 Summary

- Large numbers and small numbers are written more conveniently in a format that takes advantage of exponents: **scientific notation**.
- Normal rules of exponents apply to calculations involving multiplication or division.
- In order to add or subtract numbers in exponential notation, both numbers must be represented in the same power of ten.
- Note that numbers like 10^6 are entered in your calculator as {**1** EE **6**}, *not* {**10** EE **6**}. Other examples are 10^{-30} {**1** EE **(−)** **30**} and -10^{-12} {**(−)** **1** EE **(−)** **12**}.
- Rules of significant figures (Chapter 7) must be applied to correctly represent answers to calculations.

Practice Exercises Chapter 6

1. Convert the following to scientific notation:

    a. 54,000.00

    c. −0.00131

    b. 142.35

    d. 0.00000004

2. Convert the following from scientific notation:

    a. 1.6×10^{-4}

    c. 3.7542×10^3

    b. 0.2×10^3

    d. 4.0×10^6

3. Perform these calculations by hand, and verify your answers with your calculator:

    a. $(3.4 \times 10^{-4})(1.7 \times 10^3)$

    c. $(0.2 \times 10^{-4})(0.4 \times 10^{-4})$

    b. $(1.2 \times 10^7)(7.9 \times 10^3)$

    d. $(8.4 \times 10^5)(0.5 \times 10^{-9})$

4. Perform these calculations by hand, and verify your answers with your calculator:

    a. $\dfrac{3.4 \times 10^{-4}}{1.7 \times 10^{3}}$ c. $\dfrac{5.0 \times 10^{6}}{-2.5 \times 10^{-5}}$

    b. $\dfrac{6.5 \times 10^{-4}}{0.5 \times 10^{-6}}$ d. $\dfrac{-4.8 \times 10^{-13}}{1.2 \times 10^{5}}$

5. Perform these calculations by hand, and verify your answers with your calculator:

    a. $(2.2 \times 10^{-2}) + (1.4 \times 10^{-2})$ c. $(4.53 \times 10^{2}) - (0.2 \times 10^{3})$

    b. $(5.6 \times 10^{4}) - (2.1 \times 10^{4})$ d. $(1.1 \times 10^{-2}) - (3.0 \times 10^{2})$

Significant Figures

Thank you for tuning in to Chapter Seven. In our last episode (the end of Chapter Six), you performed an addition of two numbers expressed in scientific notation:

$$213. \quad \times 10^3$$
$$\underline{+ \ 7.55 \times 10^3}$$
$$220.55 \times 10^3$$

The problem that was mentioned earlier involves the number of digits used in representing the sum—each number to be added has three digits, yet the sum is written with five. Or is it?

One of the challenges in using numbers to describe scientific information is your confidence in the numbers. Clearly, when you count discrete items like people, you are quite confident in the number—there are 24 people in your Spanish class, not 24.2. However, when you measure a value, you are always faced with the task of representing how well you can make the measurement (the "error"). Once measurements are made and calculations are performed, you are then responsible for reporting the result in a way that reflects how any uncertainty in measurement affects the calculated answer ("propagation of error").

The correct way to propagate error involves the application of differential calculus (wait...calm down, take a deep breath and read the next sentence...). Luckily, for calculations performed in first-year science courses, it is adequate to determine error in numbers using a much-simplified system. By learning a few rules for manipulating **significant figures**, you can report numbers with the confidence expected in simple calculations. You will identify the digits in numbers that are significant and use the rules to represent results of calculations.

Rules for determining significant digits in a number:

1. Non-zero digits are always significant.

    -   Numbers like 23.7 and 4.67 each have three significant digits.

    -   Numbers like 3.467 and 112.6 each have four significant digits.

2. Zeros between significant digits (or between non-zero numbers) are significant.

    -   80.6 has three significant digits.

    -   40.06 has four significant digits.

3. Trailing zeros in the decimal portion are significant.

    -   In a number like 0.008**200**, the boldface zeros are significant.

    -   In a number like 6.2**0** × 10^4, the boldface zero is significant.

What the rules above don't tell you:

a. In a number like 0.**00**8200, the first two digits to the right of the decimal in a number less than one are NOT significant and are used as place markers for the decimal. If you write the number in scientific notation, the non-significant zeros disappear:

$$8.200 \times 10^{-3} \text{ has 4 significant digits}$$

b. The zero to the left of the decimal point in a number less than one (like **0**.008200) is NOT significant. The purpose of the zero is to prevent you from reading it as a period at the end of a sentence.

c. Trailing zeros in a whole number are not necessarily significant. For example, in the number 420, the zero might not be significant. If there are three significant digits, it should be written as 4.20×10^2.

Consider the measurement of mass. Maybe you will determine the mass of some frogs for your biology lab. Suppose you measure one frog, and it has a mass of 350 grams (about three-quarters of a pound; there are 453.6 grams per pound). How will you enter the mass of the frog in your lab notebook? It depends on how precise your measurement was—how carefully you read the balance and the sensitivity of the balance. If you used a sensitive balance that measures to the nearest hundredth of a gram (0.01 g), you could report 350.00 g; this number implies that the two digits after the decimal are important. Your uncertainty is in the last digit that you report (± 0.01 g). This is the simplified method for reporting uncertainty in measurements—when you write a number, assume that your uncertainty is in the last significant digit. Next, count all of the digits in the number. In this case, the number is 350.00, and there are five significant digits.

Let us return to your hefty frog. What if the balance only read to the nearest ten grams? Reporting the number as 350 grams then becomes incorrect. Remember that when you write a number, it is assumed that your uncertainty is in the last digit on the right. But the number 350 implies that your error is in the one's place. This is clearly wrong if the balance could only weigh items to the nearest ten grams. Here again you are saved by using scientific notation. If your uncertainty is in the ten's place, then the digit that represents the tens must be the last digit on the right: 3.5×10^2. As you can see, the 5 is the last significant digit, and it implies that the measurement was good to the closest ten grams. You would report the mass of the frog with two significant digits (the 3 and the 5).

Before showing how to manipulate numbers using proper significant digits, we need to discuss the issue of rounding. Often when you make a calculation, you will need to round the result to the proper number of significant figures. Your calculator is no help in this regard—it keeps all digits whether they are significant or not. It is up to you to round correctly so that the result you report has the correct uncertainty. The simplest rounding rules are as follows:

1. If the digit in question is less than 5, drop it and all digits after it. For example, you round the number 37.649899 to 37.6 when expressing it with three significant digits.
2. If the digit in question is 5 or more, increase the digit immediately before it by one. For example, 37.6500 and 37.683 will both round to 37.7.

Now you are prepared to use significant figures so that they work for you by revealing the uncertainty of measurements in calculations. Let's turn first to the previous example that has been troubling you so much:

$$213. \quad \times 10^3$$

$$+ \ 7.55 \times 10^3$$

$$220.55 \times 10^3$$

The first number is 2.13×10^5, which has three significant digits. In order to add this number to 7.55×10^3, it is first necessary to get **both** numbers into the *same power of ten*. This is the rule for addition and subtraction. Once in the same power of ten, the numbers can be added, but the answer must be reported with the proper number of significant figures (**sig figs** for short). If your first number had no digits to the right of the decimal, then you cannot justify an answer that does—to do so would improve your precision artificially! Therefore, the answer must be reported with the same number of digits past the decimal as the least number of digits past the decimal in the numbers being added or subtracted. That makes the result 220.55×10^3 incorrect. The proper answer would have *no digits* past the decimal: $221. \times 10^3$ (notice the rounding). The number 221. (with the decimal and no digit following) is correct, but it looks a little funny. It is preferable to rewrite the number as 2.21×10^5. One good reason not to leave it as 221. is that the decimal might be mistaken for a period at the end of a sentence, or debris left by frogs hopping on your lab notebook.

When **adding or subtracting numbers**, the answer must be reported with the same number of digits past the decimal as the **least** number of digits that appear past the decimal in the numbers being added or subtracted:

- $(4.53 \times 10^2) - (0.20 \times 10^3) = (4.53 \times 10^2) - (2.0 \times 10^2) = 2.5 \times 10^2$

  (2.0 has one sig fig after the decimal)

- $23.0252 - 17.30 = 5.73$ (17.30 has two sig figs after the decimal)

The rules for reporting the result of a multiplication or division calculation are somewhat simpler. When you perform the operation on two numbers, the result must have the same number of significant digits as the number with the **least** total significant digits. Say you measure a box with dimensions 6.4 inches by 4.55 inches by 2.35 inches. Your calculator spits out:

$$6.4 \text{ in} \times 4.55 \text{ in} \times 2.35 \text{ in} = 68.432 \text{ in}^3$$

Clearly, you can't use all five of those digits for the volume. Your biggest uncertainty was in the first measurement of 6.4 inches. That number has only two significant digits—and so must your result. Therefore, the correct volume of the box is $6.8 \times 10^1 \text{ in}^3$. Notice that 68 in^3 is correct, but since there is no decimal it is not exactly clear where the uncertainty lies.

When you **multiply or divide** two numbers, the result must have the same number of significant digits as the number with the **least** total significant digits:

- $(16.0)(3.631) = 58.1$ (16.0 has three sig figs)
- $(-0.060081)(5.0 \times 10^3) = -3.0 \times 10^2$ (5.0 has two sig figs)

One rule of significant figures that seems to get lost in the shuffle is how to treat numbers involving **logarithms**. The rule is very simple, but students seem to forget it if they are ever even exposed to it (many instructors are guilty as well!). A practical example involves calculating the pH of a solution. The pH scale is logarithmic and gives a quick indication of whether a solution is acidic (pH < 7) or basic (pH > 7). You might notice pH values in descriptions of health and beauty aids like shampoos. The pH is determined by taking the negative log of the hydrogen ion concentration (hydrogen ions, H^+, are responsible for acidity):

$$pH = -\log[H^+]$$

If the hydrogen ion concentration is 3.5×10^{-3} M (the unit of concentration is M, molarity), the pH is:

$$pH = -\log[3.5 \times 10^{-3}] = -(-2.455931956) = 2.46$$

This is the approximate pH of some soft drinks. Notice that the pH is reported with the number of digits to the right of the decimal (two) **equal** to the number of sig figs in the original number (two). Significant figures for antilog calculations are treated the same way. The antilog of −2.46 will have two significant digits since there are two digits to the right of the decimal:

$$antilog(-2.46) = 10^{-2.46} = 0.0034673685 = 3.5 \times 10^{-3}$$

When using **logarithms**, report the answer with the same number of digits to the right of the decimal as the number of sig figs in the original number:

- $\ln(622.7) = 6.4341$
- $\log(4.7) = 0.67$

The inverse, **antilog**, will have the same number of sig figs as the digits to the right of the decimal:

- $antilog(1.5) = 10^{1.5} = 0.3 \times 10^2$
- $antilog(3.71) = 10^{3.71} = 5.1 \times 10^3$

Chapter 7 Summary

- Discrete items (7 days in a week—exactly) are "perfect" numbers.
- Assume that uncertainty is in the **last** significant digit of a number.

- Trailing zeros to the **right** of the decimal are significant: 2.0000 has five sig figs.
- Zeros between non-zero digits are significant: 30.04 has four sig figs.
- In numbers less than one, the zero to the left of the decimal and to the right of the decimal before the first non-zero digit are not significant: **0.000000**40 has only two significant figures, with the non-significant figures in boldface.
- Use scientific notation to properly represent numbers with the correct number of significant digits.
- Round numbers up when the digit is 5 or greater, down for those less than 5. For example, 246.552 rounds to 246.6 and 3449.2 rounds to 3.4×10^{3}.
- When adding or subtracting, it is necessary to get **both** numbers into the *same power of ten.*
- The result from an addition or subtraction calculation has the same number of digits past the decimal as the **least** number of digits past the decimal that appear in the numbers being added or subtracted:

$$2.45677 + 9.1 = 11.6$$

- When you multiply or divide two numbers, the result must have the same number of significant digits as the number with the **least** total significant digits:

$$0.0056 \times 345.56 = 1.9$$

- When taking logarithms, the number of digits to the right of the decimal in the result equals the number of sig figs in the original number:

$$\log(350.0) = 2.5441$$

- When taking an antilog, the number of significant figures in the answer is the number of decimal places in the original number:

$$\text{antilog}(1.73) = 5.4 \times 10^{1}$$

Practice Exercises Chapter 7

1. How many significant figures are in these numbers?

   a. 20.043

   b. 3.630×10^4

   c. 0.004300

   d. 100.0010

   e. 4.06×10^{-3}

   f. 0.0204

2. Round these numbers to three significant digits:

   a. 2.74450

   b. 532.62

   c. 0.13251

   d. 4223.0

3. Round these numbers to two significant digits:

   a. 2.74450

   b. 532.62

   c. 0.13251

   d. 4223.0

4. Express the sum or difference using the correct number of significant figures:

   a. $78.834 - 78.8$

   b. $(2.2 \times 10^{-2}) + (1.4 \times 10^{-2})$

   c. $(4.213 \times 10^3) - (3.7 \times 10^3)$

   d. $320.02 - 282.4$

   e. $(5.23 \times 10^1) + (3.4 \times 10^{-1})$

   f. $8.333 + \sqrt{7}$

5. Express the product or quotient using the correct number of significant figures:

   a. $(3.33)(6.4290)$

   b. $(7.00)(8.3)$

   c. $\dfrac{4.0 \times 10^2}{0.2}$

   d. $\dfrac{0.03303}{454.34}$

   e. $(-0.05009)(1.2 \times 10^8)$

   f. $\dfrac{32.00}{0.004050}$

6. Express the logarithm or antilogarithm using the correct number of significant figures:

   a. $\ln(321.2)$

   b. $\text{antilog}(-3.121)$

   c. $-\log(2.02 \times 10^{-2})$

   d. $\text{antilog}(0.03)$

Equations

If numbers are words in the language of science, then equations are the sentences. Sorry, that was a little hokey (but true). Equations express the relationships that are valuable in quantitative descriptions of the natural world, allowing us to state explicitly how certain parameters are linked and how to use these links to predict the behavior of systems as small as an atom and as large as the universe. Good equations make for a good scientific conversation.

Sentences are fairly predictable; they contain subjects, verbs, and objects. Likewise, equations have a structure that you can always expect to see. From the simplest viewpoint, equations have a left side, a right side, and an equals sign between the two (in some cases, to be discussed later, the equals sign is replaced by an inequality). The equals sign is the heart of the equation, and the left side and right side are the same (equal!), even though they are expressed using different numbers or symbols. Let's explore a simple example to introduce components of an equation.

Suppose you have a part-time job cleaning mouse cages for a genetics project in the biology department, hauling in a steady \$5.30 per hour. You need \$250 for a new DVD player. How many hours do you need to sling mouse mess to come up with the cash? The \$250 is a **constant**; it is a number that *does not change in value*. Your hourly rate is also a constant. The only thing that changes is the number of hours that you work—it is called a **variable**. You can write down the relationship in an equation:

$$\$250 = (\$5.30 \text{ per hour}) \times (\text{hours worked})$$

To get a more compact equation you can replace "hours worked" with an algebraic variable, x, for instance. The equation becomes (dropping the units of dollars)

$$250 = 5.30x$$

The preceding equation is the simplest type, containing only one variable and two constants. Because the variable is raised to the first power, it is called a **linear equation**. Simply divide both sides of the equation by the number next to the variable (called the **coefficient**) to solve the equation:

$$\frac{250}{5.30} = \frac{5.30x}{5.30}$$
$$47.2 = x, \text{ or } x = 47.2$$

You will have to work 47.2 hours to afford the new DVD player.

You can make this equation more general by assigning a variable to your total salary. The following equation describes your total salary (S) as a function of your hourly rate (5.30) and the number of hours worked (x):

$$S = 5.30x$$

Here S takes the place of your total salary, which depends on your hourly rate (a constant in this example) and the number of hours that you work (a variable). Using this equation, you can determine how much you will earn for any number of hours worked; simply replace x with the number of hours on the job.

There are several ways your graphing calculator can show a linear relationship. Type the equation under {Y = } and go to {2nd TABLSET} on the top row of keys. Change TblStart to 0 and ΔTbl to 1, then generate a table with {2nd TABLE} (over the {GRAPH} key at the top).

This creates a table that shows how much money you will make for every hour worked. You can use the down arrow key to reveal more hours worked. In the preceding center screen display, you instructed your calculator to start the table with $x = 0$ and to increment the hours worked by 1 (TblStart and ΔTbl). How much money will you make after working 40 hours?

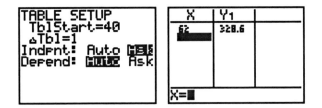

If you want to type your own values into the calculator, press {2nd TBLSET} again, but this time change the independent variable from Auto to Ask. When you go to the table you may type in any value of x you desire.

In an earlier example, you were asked how long it would take to earn $250. You could graph the equation with $y_2 = 250$, since the y-coordinate is the amount of money earned. Change the window so that the y-max includes 250, as shown:

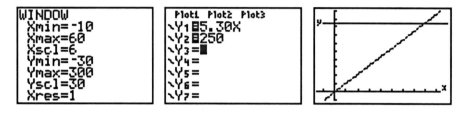

Now go to {2nd **CALC** 5:**intersect**} to find the corresponding number of hours worked for $250 earned. When the calculator prompts for a guess, just press {**ENTER**}.

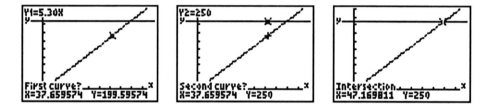

The key to using equations is rearranging them to give the value that you seek. Rearranging equations is a skill that you need to develop if you intend to be able to solve problems in science. It is simpler than it appears, and with some practice you will be able to rearrange equations quickly and see relationships that are important to your understanding of a subject. Consider the bottom line: You can only do to one side of an equation *what you do to the other*.

Most equations will require manipulation for you to solve them when doing problems. Once you have identified the correct equation (a job in itself) you must rearrange it properly so that you can input numbers and perform a calculation for the answer. In practice you will be presented with information (data) to use in an equation that you will manipulate if the situation calls for it. By moving variables around, you will rearrange the equation so that you can determine an answer to the problem. Remember, you can only solve for *one variable* with a single equation—you must have a value for **all** other variables and constants. If you have an equation with two unknown variables, you will need another equation that relates the two.

We'll look at this again in Chapter 12, Solving Problems. Right now let's look at other examples of equations and how to rearrange them.

Consider Coulomb's Law, which describes the force F between two particles of charge q and Q separated by distance r, where k is a constant:

$$F = \frac{kqQ}{r^2}$$

Suppose you needed to solve for r. You should recognize that to get r alone you must first get r^2 alone. This is accomplished by multiplying both sides by r^2 (or cross multiplying as done in Chapter 3) to yield

$$r^2 F = kqQ$$

and dividing both sides by F to give

$$r^2 = \frac{kqQ}{F}$$

Now the value of r can be determined by taking the square root of **both sides** of the equation:

$$\sqrt{r^2} = \sqrt{\frac{kqQ}{F}}$$

which yields

$$r = \sqrt{\frac{kqQ}{F}}$$

The previous examples involve variables and constants that are only multiplied or divided. Often equations involve terms that are added or subtracted. Remember, when rearranging an equation you must do the same mathematical manipulation to **each** side. So if you subtract a number from the left side, you must also subtract the same number from the right side. For example, the following equation relates the Fahrenheit and Celsius temperature scales:

$$T_F = \left(\frac{9\ °F}{5\ °C}\right)T_C + 32\ °F$$

To solve for the temperature in Celsius, T_C, first subtract 32 °F from each side,

$$T_F - 32\ °F = \left(\frac{9\ °F}{5\ °C}\right) + 32\ °F - 32\ °F$$

$$T_F - 32\ °F = \left(\frac{9\ °F}{5\ °C}\right)T_c$$

(note how the 32 °F moves to the left side and becomes **negative**), then multiply each side by the reciprocal of the conversion constant

$$\left(\frac{5\ °C}{9\ °F}\right)(T_F - 32\ °F) = \left(\frac{5\ °C}{9\ °F}\right)\left(\frac{9\ °F}{5\ °C}\right)T_C$$

and reverse the equation to isolate T_C:

$$T_C = \left(\frac{5\ °C}{9\ °F}\right)(T_F - 32\ °F)$$

So much for the easy stuff. Let's look at an equation that combines the challenges of multiplication and division with addition and subtraction. For a sample of gas, the van der Waals equation of state relates pressure (P), volume (V), number of particles (n, the number of moles), and temperature (T). There are two van der Waals constants, a and b, which are different for different gases. This equation is an improvement on the "ideal" gas law (which contains the gas constant, R),

$$PV = nRT$$

where the variables P and V have been expanded using the van der Waals constants:

$$\left(P + \frac{an^2}{V^2}\right)(V - nb) = nRT$$

To rearrange this equation to solve for P, first divide both sides by (V-nb):

$$\frac{\left(P + \dfrac{an^2}{V^2}\right)(V - nb)}{(V - nb)} = \frac{nRT}{(V - nb)}$$

$$\left(P + \frac{an^2}{V^2}\right) = \frac{nRT}{(V - nb)}$$

To isolate P, you must move the term containing the constant a by subtracting the term $\dfrac{an^2}{V^2}$ from both sides:

$$\left(P + \frac{an^2}{V^2}\right) - \frac{an^2}{V^2} = \frac{nRT}{(V - nb)} - \frac{an^2}{V^2}$$

$$P = \frac{nRT}{(V - nb)} - \frac{an^2}{V^2}$$

There, that wasn't so bad, was it? Now, for practice, solve for the constant b.

SPECIAL EQUATIONS

There are some special formats for equations that will come up often in problems that you solve. One such equation is a **linear equation**. In a linear equation, the highest power of the variable is 1 (so the degree of the equation is 1), such as

$$4x - 8 = 2x + 4$$

To solve a linear equation, put the variable terms on one side and the constant terms on the other. In other words, to get the variables on one side, subtract $2x$ from both sides:

$$4x - 8 - 2x = 2x + 4 - 2x$$
$$2x - 8 = 4$$

To put the constants on the other side, add 8 to both sides:

$$2x - 8 + 8 = 4 + 8$$
$$2x = 12$$

To solve for x, divide both sides by 2:

$$\frac{2x}{2} = \frac{12}{2}$$
$$x = 6$$

To check your answer, substitute 6 for x into the original equation:

$$4(6) - 8 = 2(6) + 4$$
$$24 - 8 = 12 + 4$$

Since this is a true statement, you solved the equation correctly.

Another common equation is a **quadratic equation** in which the highest power of the variable is 2 (or degree 2); the general form for a quadratic equation is

$$ax^2 + bx + c = 0$$

You can solve a quadratic several ways, including algebraically, graphically, and by using the quadratic formula. For example, to solve this quadratic equation

$$5x - 2 = 2x^2$$

use algebra to manipulate the equation so that one side of the equation equals 0. Move the terms from the left side of the equation to the right because it will be easier to solve when the highest degree term (the one that has the variable squared) is positive. In order for the terms to cross the equals sign, you must subtract $5x$ and add 2 to both sides, arranging the terms so that the highest power of the variable is first, all the way down to the constant term:

$$0 = 2x^2 - 5x + 2$$

This equation can be factored as follows:

$$0 = (2x - 1)(x - 2)$$

Before going on, check your factoring by multiplying it out. One method for doing this is called **FOIL**, a mnemonic device that stands for First, Outer, Inner, and Last. You will multiply the terms out in this order (**FOIL**) as shown:

$$(2x)(x) + (2x)(-2) + (-1)(x) + (-1)(-2) = 0$$

$$2x^2 - 4x - x + 2 = 0$$

$$2x^2 - 5x + 2 = 0$$

Since the factoring is correct, let's go back to the factored form

$$(2x - 1)(x - 2) = 0$$

To finish, set each factor to 0, because if the product of two numbers is 0, at least one of the numbers would have to be zero:

$$2x - 1 = 0 \ \text{ or } \ x - 2 = 0$$

$$x = \frac{1}{2} \quad \text{or} \quad x = 2$$

Another way to solve this quadratic equation is to graph it on your calculator. Simply type in the equation under {**Y =** }, and see where the graph crosses the x-axis as shown:

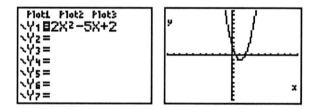

Use the calculator to determine the solutions (also called **zeros**) by using the {**CALC**} key above the {**TRACE**} key. Press {**2ⁿᵈ CALC 2:zero ENTER**}. For a left bound (lower x value), move the cursor above the x-axis, to the left of $x = \frac{1}{2}$, and press {**ENTER**}. For a right bound (higher x value), move the cursor below the x-axis, to the right of $x = \frac{1}{2}$, and press {**ENTER**}. Ignore the GUESS prompt and press {**ENTER**} one more time. The calculator will display the zero. Here is the display to find the zero $x = 2$:

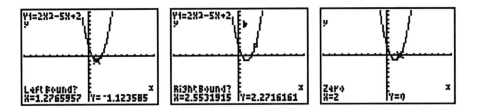

Quadratic equations can be solved using the quadratic formula. Recall that all quadratic equations can be written in the form

$$ax^2 + bx + c = 0$$

Use the coefficients a, b, and c to solve the equation with the **quadratic formula**:

$$x = \frac{-b \pm \sqrt{b^2 - 4ac}}{2a}$$

Note that there may be two values of x, since the ± symbol instructs you to do the calculation two ways—one by adding the term in the square root symbol and the other by subtracting it. So for our example,

$$a = 2 \quad b = -5 \quad c = 2$$

and,

$$x = \frac{-(-5) \pm \sqrt{(-5)^2 - 4(2)(2)}}{2(2)}$$

will give the following solutions:

$$x = \frac{5 \pm \sqrt{25 - 16}}{4} = \frac{5 \pm \sqrt{9}}{4} = \frac{5 \pm 3}{4} = 2, \tfrac{1}{2}$$

The quadratic formula gave the same solutions that we determined from the factored form. Just remember to set your quadratic equation equal to **zero** before factoring, graphing, or using the quadratic formula. Also keep in mind that there may not be two solutions. There could be one solution (if the graph just touches the x-axis one time), or there may be *no* solution (if the graph doesn't cross the x-axis at all). Here is a quadratic equation that only has one solution:

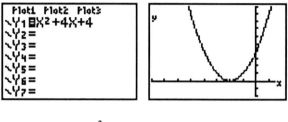

$$x^2 + 4x + 4 = 0$$

To solve algebraically, factor:

$$(x + 2)(x + 2) = 0$$
$$x = -2$$

This example does not cross the x-axis:

$$3x^2 + x + 3 = 0$$

$$x = \frac{-1 \pm \sqrt{1^2 - 4(3)(3)}}{6} = \frac{-1 \pm \sqrt{1 - 36}}{6} = \frac{-1 \pm \sqrt{-35}}{6} = \frac{-1 \pm i\sqrt{35}}{6}$$

This quadratic equation will not factor. If you use the quadratic formula, you will get a negative number under the square root, which means the solutions are imaginary. Imaginary solutions mean the graph will not cross the *x*-axis!

INEQUALITIES

Some problems will require that you not equate things but rather set limits using **inequalities**. You might encounter the inequalities **less than**, <, **less than or equal to**, ≤, **greater than**, >, or **greater than or equal to**, ≥. You should treat equations that contain inequalities *the same* as equations with an equals sign, except in one very important way. When you multiply or divide your equation by a **negative** number, you must **reverse** the direction of the inequality (< becomes >, and > becomes <, for instance). A simple example will illustrate the point. The following inequality is true:

$$-2 < +5$$

If you were to multiply this equation by –3, you must reverse the inequality to keep the equation true:

$$(-3)(-2) \ ? \ (-3)(+5)$$
$$+6 > -15$$

The preceding treatment of inequalities is the same for equations that contain variables. As an example, let's solve for *x* in the following equation. As in a linear equation, you want to place variable terms on one side and constant terms on the other:

$3x - 7$	>	$5x + 9$	original equation
$-2x$	>	16	subtract 5*x* from both sides; add 7 to both sides
$\dfrac{-2x}{-2}$	<	$\dfrac{16}{-2}$	divide both sides by –2, reverse the inequality
x	<	-8	simplify

The **only** time you switch the inequality sign is when you multiply or divide by a negative number; in any other case, leave the inequality sign alone!

Chapter 8 Summary

- Equations have a left side, a right side, and an equals or inequality sign between the two.
- Components of equations can be **variables** (things that change), **coefficients** (numbers multiplied by the variables), or **constants** (numbers that do not change).
- You can only do to one side of an equation *what you do to the other*.
- The general form for a quadratic equation is $ax^2 + bx + c = 0$. The right side must be equal to **zero** to solve any quadratic equation.
- To solve a quadratic on the calculator, graph the quadratic and use the CALC function on the calculator to find the zeros.
- The **quadratic formula** is

$$x = \frac{-b \pm \sqrt{b^2 - 4ac}}{2a}$$

- Equations can contain the inequalities less than, <, less than or equal to, ≤, greater than, >, or greater than or equal to, ≥.
- When multiplying or dividing an inequality by a **negative** number, you must **reverse** the direction of the inequality.

Practice Exercises Chapter 8

1. Solve the following equations for the variable:

   a. $-9r + 8 = -91$

   b. $15x + 20 = 8x - 22$

   c. $9y - 7y = 48$

   d. $5 - 4a = a - 13$

2. Solve each for y in terms of x:

    a. $2x + 2y = x^2$
                  d. $2\ln x + \ln y = 2$

    b. $\dfrac{y+1}{x+1} = 2$
                e. $E = RT\ln\left(\dfrac{1}{y}\right)$

    c. $\ln x > 2 - y$
                  f. $xy + y - 7 = 0$

3. Solve each inequality:

    a. $3x + 5x < 16$
                  c. $5y \le \dfrac{3}{2}$

    b. $6 - 5y \ge -9$
                 d. $17 - 5y < 8y - 5$

4. Solve each formula for the given variable:

    a. $d = rt$, solve for r
              c. $A = \dfrac{1}{2}bh$, solve for b

    b. $P = 2(l + w)$, solve for w
        d. $C = \dfrac{Wc}{1000}$, solve for c

5. Find the real zeros of each quadratic equation by:

       i.   factoring (if possible)
      ii. using the quadratic formula
     iii. graphing on your graphing calculator

    a. $x^2 - 2x + 1 = 16$
              d. $x^2 - 4x - 11 = 0$
    b. $6x^2 + x - 1 = 0$
              e. $x^2 - 4x + 11 = 0$
    c. $3x^2 - 10x = 8$
               f. $2x^2 + 4x = 0$

UNITS

Units give value to numbers. They tell us what measurement has been made and in what magnitude. Consider, for example, distance: the number "5" does not convey enough information about a distance. Is it 5 centimeters, or 5 kilometers? There is a big difference. What about the differences between **m**, **mm**, **cm**, and **km**? Prefixes of units can relate orders of magnitude. Knowing units, their prefixes, and how to use them will help solidify your understanding of numbers, equations, and the measurements used in science.

Units fall into one of two categories, the *fundamental* units and the *derived* units. The fundamental units include length, mass, time, and temperature. Derived units are combinations of fundamental units. For example, velocity is found from distance (length) divided by time, so velocity has units of length divided by time:

$$\text{velocity} = \frac{\text{length (meters)}}{\text{time (seconds)}} = \frac{\text{m}}{\text{s}}$$

Another unit of velocity would be miles per hour, again a length (miles) divided by time (hours). All derived units involve application of an equation using the proper fundamental units (or other derived units). That is the power of using units: if you manipulate units properly, the answer you

Table 9-1 SI Base Units

Quantity	Unit	Symbol
Length	meter	m
Mass	kilogram	kg
Time	second	s
Temperature	kelvin	K

determine from a formula will have the right units. Always "carry" your units—they will guide you to the right answer. Once your computation is done, if the result has the wrong unit, then it will raise a red flag. You can detect mistakes by analyzing why your calculated unit is in error.

Scientists use units of the International System of Units (Système International, or **SI**). Table 9-1 lists **SI** base units used around the world. There are also standardized prefixes used with units (those listed in Table 9-1) and derived units as well. These prefixes represent the order of magnitude by which the unit is multiplied. For instance, 1000 grams (1000 g) is a kilogram (1 kg), since **kilo** is the prefix for 1000. These SI prefixes are listed in Table 9-2. The most commonly used prefixes are **nano**, **micro**, **milli**, **centi**, **kilo**, **mega**, and **giga**.

Converting units is a skill that you must master in order to solve problems correctly. Conversions are of three types: conversions that involve orders of magnitude (1000 mm per meter), conversions that involve definitions (60 s per minute), and conversions between measurement systems (2.54 cm per inch). The most typical system conversions are between the **Metric System** and the **U.S. Customary Units System** (commonly called the English System). Instead of trying to remember the limitless number of conversion factors, just focus on two that will allow conversion of length and mass: 2.54 cm/in and 453.6 g/lb. The international committee sets the length conversion, and it is a "perfect" conversion—it has as many significant figures as you need. Strictly speaking, the second conversion is only true if you make your measurements under the Earth's constant gravitational field, since pounds (lb) are a measure of force, not mass. If you can remember these two

Table 9-2 SI Prefixes

Prefix	Symbol	Value
atto	a	10^{-18}
femto	f	10^{-15}
pico	p	10^{-12}
nano	**n**	$\mathbf{10^{-9}}$
micro	**μ**	$\mathbf{10^{-6}}$
milli	**m**	$\mathbf{10^{-3}}$
centi	**c**	$\mathbf{10^{-2}}$
deci	d	10^{-1}
deka	da	10^{1}
hecto	h	10^{2}
kilo	**k**	$\mathbf{10^{3}}$
mega	**M**	$\mathbf{10^{6}}$
giga	**G**	$\mathbf{10^{9}}$
tera	T	10^{12}
peta	P	10^{15}
exa	E	10^{18}

conversions, and convert your numbers to cm/in or g/lb where appropriate, you will always be able to make the proper conversion.

A conversion is really a ratio that has a value equal to *one* (you can always multiply by the number one). Multiply the number you want to convert by the conversion expressed with the proper combination of units. The following example shows all three types of conversions. Note that in each step you cancel one unit by multiplying or dividing by the proper conversion. The result is the same number expressed using the new unit.

The Earth is about 144 billion meters from the sun. How far is this in miles?

$$\left(144 \times 10^9 \text{ m}\right)\left(\frac{100 \text{ cm}}{1 \text{ m}}\right) = 144 \times 10^{11} \text{ cm} \qquad \text{[order of magnitude change]}$$

$$\left(144 \times 10^{11} \text{ cm}\right)\left(\tfrac{1 \text{ in}}{2.54 \text{ cm}}\right) = 5.67 \times 10^{12} \text{ in} \quad \text{[system conversion]}$$

$$\left(5.67 \times 10^{12} \text{ in}\right)\left(\tfrac{1 \text{ ft}}{12 \text{ in}}\right) = 4.72 \times 10^{11} \text{ ft} \qquad \text{[inches to feet, should know]}$$

$$\left(4.72 \times 10^{11} \text{ ft}\right)\left(\tfrac{1 \text{ mi}}{5280 \text{ ft}}\right) = 89.5 \times 10^{6} \text{ mi} \qquad \text{[feet to miles, handy to know]}$$

So the Earth is about 90 million miles from the sun. The strategy in the previous example was to convert meters to centimeters so that you could use your centimeters-to-inches conversion factor. Once the number is in inches you can get to miles using the definitions of feet and miles. The conversion took a few steps, but it is logical and easy to do—or at least easier than remembering that there are 1609 meters per mile! You can combine the conversions into one big equation:

$$\left(144 \times 10^{9} \text{ m}\right)\left(\frac{100 \text{ cm}}{1 \text{ m}}\right)\left(\frac{1 \text{ in}}{2.54 \text{ cm}}\right)\left(\frac{1 \text{ ft}}{12 \text{ in}}\right)\left(\frac{1 \text{ mi}}{5280 \text{ ft}}\right) = 89.5 \times 10^{6} \text{ mi}$$

or

$$\frac{(144 \times 10^{9} \text{ m})(100 \text{ }^{cm}/_{m})}{(2.54 \text{ }^{cm}/_{in})(12 \text{ }^{in}/_{ft})(5280 \text{ }^{ft}/_{mi})} = 89.5 \times 10^{6} \text{ mi}$$

The advantage of using one conversion equation is that you can see more clearly how various units cancel to yield a number in the desired unit. Units that appear in both the denominator and numerator will cancel. The drawback is that you have to be careful to avoid losing your place when making decisions about whether to put a conversion in the numerator or the denominator. Notice in the preceding example that the conversion factors are exact numbers and do not affect the multiplication and division rules of significant figures. The use of 12 in/ft does *not* make the final answer 9.0×10^{7} mi.

In the previous example, conversions were expressed in fractional form. Sometimes you may encounter units written using an in-line format. Negative exponents are used for units that appear in the denominator of derived units. In this alternative format, 2.54 cm/in becomes $2.54 \text{ cm} \cdot \text{in}^{-1}$. Force is measured in the derived unit *newtons*, N. Recall that force is equal to mass (kg) times acceleration (m/s²):

$$F = m(\text{kg}) \cdot a\left(\frac{m}{s^2}\right) = (m \cdot a)\frac{\text{kg} \cdot \text{m}}{\text{s}^2}$$

Using the in-line format, newtons can be expressed as $\text{kg} \cdot \text{m} \cdot \text{s}^{-2}$. The raised periods represent multiplication of the fundamental units. Learn to recognize the in-line format by using it when you do practice exercises. The fractional format may be easier for you to use, so you will need to convert freely between the two formats.

One pitfall to watch out for is the use of conversion factors that are raised to powers. Consider a 10 ft × 10 ft × 8.5 ft room with a volume of 850 ft³. When you convert this volume to cubic meters, you must also cube the conversion factors:

$$850 \text{ ft}^3 \times \left(\frac{12 \text{ in}}{1 \text{ ft}}\right)^3 \times \left(\frac{2.54 \text{ cm}}{1 \text{ in}}\right)^3 \times \left(\frac{1 \text{ m}}{100 \text{ cm}}\right)^3 =$$

$$850 \text{ ft}^3 \times \left(\frac{1728 \text{ in}^3}{1 \text{ ft}^3}\right) \times \left(\frac{16.4 \text{ cm}^3}{1 \text{ in}^3}\right) \times \left(\frac{1 \text{ m}^3}{10^6 \text{ cm}^3}\right) = 24 \text{ m}^3$$

Chapter 9 Summary

- Units are important—they indicate what measurement has been made and in what magnitude.
- Units fall into two categories, *fundamental* units and *derived* units.
- Always "carry" your units, they will guide you to the right answer. By canceling units as you use an equation, you will avoid mistakes.
- Learn the **SI** base units and standardized prefixes.
- A unit conversion is really a ratio that has a value equal to *one*.
- Commit the conversions 2.54 cm/in and 453.6 g/lb to memory.

- Combine conversions into one big equation for clarity.
- Cancel units that appear in both the numerator and denominator.
- Units are sometimes written in an in-line format using negative exponents for units that appear in the denominator.

Practice Exercises Chapter 9

1. Change these numbers to equivalents using unit prefixes (for example, 1000 g = 1 kg):

   a. $5,000,000

   b. 2,000,000,000 bytes

   c. 0.025 m

   d. 4×10^{-9} s

   e. 8×10^{-3} W

   f. 10^{-6} m

   g. 50×10^{3} tons

   h. 1×10^{12} dactyls

2. Rewrite these numbers without using the unit prefix:

   a. 5.3 Mbar

   b. 0.2 kJ

   c. 20 mm

   d. 1.1 ng

3. Perform these unit conversions:

   a. 1.0 yr to s

   b. 5.29×10^{-11} m to pm

   c. 500 yd to mi

   d. 0.0025 mi to m

   e. 0.35 lb to kg

   f. 1.2 μs to ns

   g. 101.3 MHz to KHz

   h. 100 g to lb

4. Change these units to the in-line format:

   a. $96485 \ \frac{C}{mol}$

   b. $0.08206 \ \frac{atm \cdot L}{mol \cdot K}$

   c. $2.998 \times 10^{8} \ \frac{m}{s}$

   d. $8.85 \times 10^{-12} \ \frac{C^2}{N \cdot m^2}$

5. Change these units from in-line format to fractional format:

   a. $1 \, J \cdot s^{-1}$

   b. $9.81 \ m \cdot s^{-2}$

   c. $8.314 \, J \cdot mol^{-1} \cdot K^{-1}$

   d. $1 \, kg \cdot m^2 \cdot s^{-2}$

Graphing

Graphs are pictures that express mathematical relationships. You are probably already familiar with many types of graphs since they are appearing more often in the print media and on television. Political candidates, for example, use graphs to try to sell their economic plans on TV. The most common graphs used in the media are the **bar chart** and the **pie chart**. Pie charts are circles divided into segments; each segment represents a fraction of the whole contributed by a certain characteristic. In a recent poll, 1000 chemistry students were asked to indicate their gender for statistical purposes as part of a survey. The results are presented in Figure 10-1 in the form of a pie chart. What is your interpretation of this data?

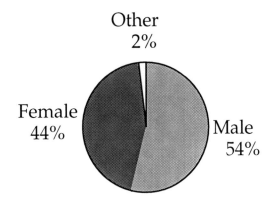

Figure 10-1   Example of a pie chart.

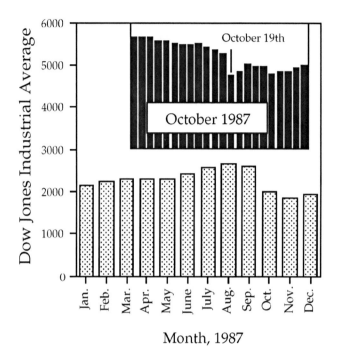

Figure 10-2 Bar chart of the Dow Jones Industrial Average for the year 1987.

The bar chart differs from the pie chart. In a pie chart the sum of all segments must equal 100%. However, the bar chart represents the behavior of one parameter as a function of another. A common use of bar charts is to show how some characteristic varies with time. Evening news programs often use graphical representation to show the fluctuations in the stock market. The length of the bar represents the Dow Jones Industrial Average (an overall gauge of stock value), and the position of the bars typically advances from left (early time) to right (later time) at regular intervals. Figure 10-2 shows such a chart for the year 1987. A close-up of October (see insert in Fig. 10-2) shows Monday, October 19th, when the Dow Jones Industrial Average tumbled almost 500 points (data from *The Dow Jones Averages© 1885-1990*, Edited by Phyllis S. Pierce, Business One Irwin, Homewood, Ill., 1991). It is easy to understand at a glance the fluctuations in stock values by noting the variation in the height of the bars (this will become considerably more important to you after graduation when you are making major bucks).

Bar charts are useful when each value along the horizontal axis (the *x*-**axis**) can have only one amount associated with it. Often this is not the case. In addition, for a given set of information (data), the value of the parameter represented along the *x*-axis (the **independent variable**) may not necessarily occur in regular intervals. We can represent the measured parameter (the **dependent variable**) as a function of the independent variable in a more flexible format if we let the resulting value be a point on a two dimensional grid. A collection of points that represents this behavior is called a **scatter plot**. Points have **coordinates** that represent values of the independent variable (*x*) and the dependent variable (*y*). The dependent variable is plotted on a vertical axis (*y*-**axis**) that is positioned perpendicular to the *x*-axis. Each point has a position determined by two parameters, *x* and *y*, so the coordinates are written in the format (*x,* *y*). Figure 10-3 shows the stock market data plotted in the scatter plot format, with lines joining points to aid the eye.

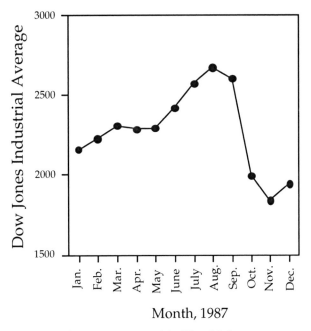

Figure 10-3   Scatter plot of data presented in Fig. 10-2.

A scatter plot is usually just the beginning in the treatment of scientific data. In constructing an experiment, you normally control one variable (time or temperature, for example) and make measurements of the dependent variable relative to it. This data becomes a collection of (x, y) pairs. Once these points are plotted, the result may follow a specific relationship. One of the most important relations is the equation of a line. The value of y behaves predictably as a function of x. The equation has a particular form, called the **slope-intercept** form of a line:

$$y = mx + b$$

where m is the **slope** of the line and b is the **y-intercept**. The y-intercept is the value of y when $x = 0$ (where the line crosses the y-axis). The slope is the tilt of the line, determined by how much y values change with respect to x values. This change is known as the "rise over run," which means that you can determine the slope by comparing two points on the line, (x_1, y_1) and (x_2, y_2):

$$m = \frac{\text{rise}}{\text{run}} = \frac{y_2 - y_1}{x_2 - x_1} = \frac{\Delta y}{\Delta x}$$

The symbol Δ means "change in" and instructs you to take a difference between two values. Two points determine a line, and the equation of the line can be determined from *any* two points on the line. The traits of a line are summarized in Figure 10-4. The slope of the line in Figure 10-4 is negative; it can be described as slanting "upward left." A line with m > 0 (positive) slopes the other way and can be described as slanting "upward right."

For example, suppose you want to find the equation of a line that passes through the two points, $(1, 2)$ and $(4, 5)$. First find the slope of the line:

$$m = \frac{y_2 - y_1}{x_2 - x_1} = \frac{5 - 2}{4 - 1} = \frac{3}{3} = 1$$

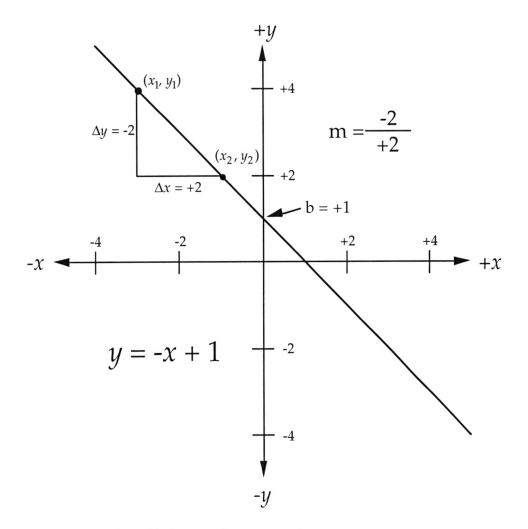

Figure 10-4   A line with the equation $y = -x + 1$.

Now substitute $m = 1$ and either of the two points into the general equation of a line ($y = mx + b$) to solve for b. Let's use the ordered pair (1, 2):

$$2 = (1)(1) + b$$
$$b = 1$$

Therefore, substituting back into $y = mx + b$, the equation of the line is

$$y = 1x + 1 \quad \text{or} \quad y = x + 1$$

This may be easier using the **point-slope** form of a line:

$$y - y_1 = m(x - x_1)$$

Here, just substitute the slope (m) and either point for (x_1, y_1):

$$y - 2 = 1(x - 1)$$
$$y = x - 1 + 2$$
$$y = x + 1$$

This line, of course, can be checked using your graphing calculator. Trace to make sure the line is passing through the original two points.

Once the equation for a line is known, all points can be determined, particularly those that lie between data points (**interpolation**) and those that lie outside of the data range (**extrapolation**). New values for y can be interpolated or extrapolated from the slope, intercept, and a given value of x by inputting all known information into the $y = mx + b$ equation:

$$y_{new} = mx_{new} + b$$

Interpolation can also be performed for situations in which the data follows a curve, if the behavior between two closely spaced points can be approximated by a line. In this instance, the two data points are used to "define" a line of interpolation, and the new point is determined in the normal fashion. This method is valuable only if the functional behavior is close to linear between the two known data points. For the same reason, care must be exercised when extrapolating values beyond the range of collected data. If the line obtains curvature outside of the data range, then an extrapolation will yield erroneous results. Only when a function is known not to deviate from linear behavior can extrapolations be performed successfully. Never stake your reputation on an extrapolation. There have been numerous examples of linear relationships that deviate at higher and lower values. In fact, these deviations themselves can become a field of study. Nonlinear effects are becoming increasingly important as we stretch our understanding of systems both ultralarge and ultrasmall.

The ease of fitting data to a straight line can be exploited with equations that are not normally linear. Often a relation is converted from its original form to a linear format so that measurements can be plotted as (x, y) pairs and a slope can be determined. One common transformation of an equation is to take the logarithm (roots are also useful in this regard). The new form of the equation is set up in the linear fashion that parallels $y = mx + b$. An obvious example is an equation that contains an exponential term:

$$k = Ae^{\frac{-E_a}{RT}}$$

This is the Arrhenius Equation that relates rate constants of chemical reactions to temperature. When the natural logarithm is applied to this equation, it becomes:

$$\ln(k) = \frac{-E_a}{R}\left(\frac{1}{T}\right) + \ln(A)$$

Values of k are converted to $\ln(k)$ and plotted versus the inverse of temperature $(1/T)$. A best fit line yields a slope of $m = -E_a/R$ and y-intercept of $\ln(A)$. Therefore, a value for the activation energy (E_a) of a biochemical reaction, for example, can be determined graphically from a plot of the natural log of rate constant k as a function of inverse temperature.

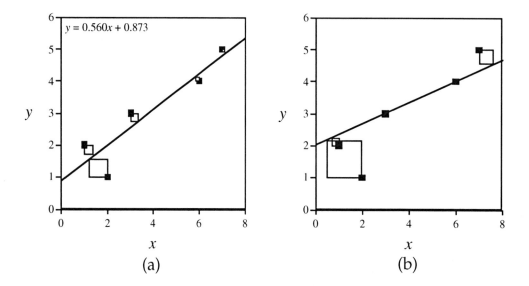

$$y = 0.560x + 0.873$$

Figure 10-5 Linear fits to sample data (Table 10-1); (a) Least Squares fit, (b) alternate line through the data (not the best fit).

Of course, it is rare that a collection of data will form a perfect straight line. The experimenter must place a line through the data that represents the "best fit" linear relation. When plotted points fall very close to a straight line, the line can be drawn by eye using a straight edge and good judgment. However, to be more exacting, there is a well-established method for determining the best-fit straight line to a body of data. This technique is called the **Least Squares Method**. If you do all of your graphing using a computer or graphing calculator, this method is but a few keystrokes away. It is valuable to understand how this method works (and where the name comes from). Figure 10-5(a) shows the best-fit straight line through a scatter plot of (x, y) data. Each data point is connected to this best line using a square. For the best-fit line, the sum of areas for all of these squares will be at a minimum. Compare the best-fit line in Figure 10-5(a) with an alternate line in Figure 10-5(b). For the best line, the sum of the areas of the squares is smaller. Hence the name Least Squares.

Table 10-1 Data for Least Squares Method

x	y	x^2	xy
1	2	1	2
2	1	4	2
3	3	9	9
6	4	36	24
7	5	49	35
Σ 19	15	99	72

As with all things that your computer can do, line fitting was done first by hand. The fitting routine that spreadsheets and plotting programs use is simple enough that you can do it by hand if there aren't too many points. The (x, y) data for Figure 10-5 is listed in Table 10-1 along with the additional information needed to compute the slope and y-intercept for the best-fit line. The symbol Σ (sigma) represents summation, adding all values of a particular type. You should be familiar with summation from taking an **average**, or **mean**, of a group of numbers. If there are N numbers with values x_i, the mean is the sum of all the numbers divided by N:

$$\text{mean}(x) = \frac{\sum_{i=1}^{N} x_i}{N}$$

The subscript i is an index for each number in the group. The summation is performed by adding all numbers, from $i = 1$ to $i = N$. This is indicated below and above the summation sign (Σ). The index below the sigma is the starting number, and the index above the sigma is where the series ends.

For **n** data points [(x, y) pairs], the Least Squares equations are:

$$\text{slope} = m = \frac{n\sum x_i y_i - \sum x_i \sum y_i}{n\sum x_i^2 - \left(\sum x_i\right)^2}$$

$$\text{intercept} = b = \frac{\sum x_i^2 \sum y_i - \sum x_i \sum x_i y_i}{n\sum x_i^2 - \left(\sum x_i\right)^2}$$

Note that the denominator for both equations is the same. The i subscripts indicate that the sums encompass all data values in the series. When you calculate xy values, you use the same point (hence the same subscript i) and multiply $x \cdot y$. Most scientific and graphing calculators have the ability to perform a Least Squares fit to (x, y) data.

We will now demonstrate how to instruct your graphing calculator to take your data, draw a scatter plot, and find an equation of best fit. Suppose you have data of how gas volume depends on pressure changes at a constant temperature, and these are your data points:

Pressure (atm)	Volume (L)
0.100	224
0.200	112
0.400	56.0
0.600	37.3
0.800	28.0
1.000	22.4

To draw a scatter plot on your calculator, press {**STAT 1:Edit ENTER**}, and enter the Pressure column data under L_1 and the Volume column data under L_2 as shown:

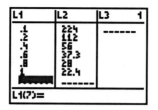

To draw the scatter plot, first press {2nd STATPLOT 1:Plot1...On ENTER}. From the six possible lists, specify that the x-coordinates are listed under L_1 and the y-coordinates are listed under L_2. To graph with an appropriate window, press {ZOOM 9:ZoomStat ENTER}.

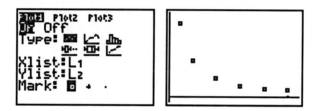

Before finding the equation of best fit, press {2nd CATALOG} and scroll down and choose DiagnosticOn (then press ENTER twice), which will show the correlation coefficient. Best-fit equations will have a correlation coefficient (**r**) very close to –1 or 1. To find the equation of best fit, press {STAT CALC} and choose from menus 4-9, 0, A, B, or C:

Scroll through the different regression equations and find the equation with the correlation coefficient (**r**) closest to 1 or –1. Note that the quadratic, cubic, and quartic regressions will only show an R^2. In our case, the equation of best fit is the Power Regression as shown:

After you press **ENTER**, type in {**VARS Y-Vars 1:Function   1:Y$_1$ ENTER ENTER**} to paste the function under Y$_1$ to graph.

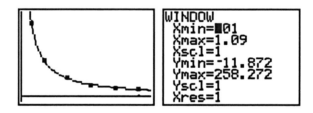

Note that **r** is very close to –1, which means the equation is a great fit! This equation is an inverse variation that occurs in the form $y = k/x$; our equation has x raised to the –1 power and can be written $y = 22.4/x$. To show the nice fit of the equation, press {**GRAPH**}.

An additional aspect of graphing experimental data involves proper treatment of error in your numbers. When a measurement is made, there is inherent uncertainty in the number, and this uncertainty is expressed graphically as an **error bar**. The bar shows the range of uncertainty in the measurements, $y \pm$ (*error*), which you can often estimate with reasonable accuracy. While the error bar is actually an "error box," since there is error in both the x and y dimensions, the error in the dependent variable is usually much greater than the error in the independent variable. For this reason, it is customary to plot error in the y-dimension only, unless the error in x is large.

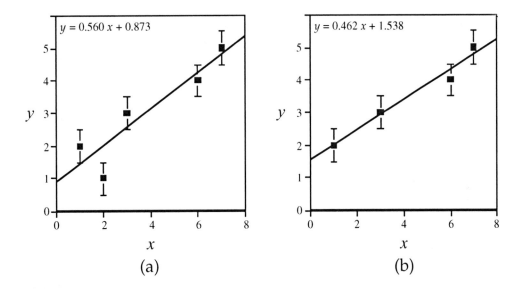

Figure 10-6   Linear fits to sample data (Table 10-1) with arbitrary error bars; (a) Least Squares fit to all points, (b) improved Least Squares fit after elimination of "bad" point at (2, 1).

Figure 10-6(a) shows the data from Table 10-1 plotted using error bars in the y-dimension (error was chosen as $\pm$ 0.5 for illustration; detailed error analysis is actually a complicated issue and will not be treated here). Note that this extra information suggests that the point at (2,1) may be erroneous, since the best-fit straight line does not pass through its error range. This result should cause the experimenter to reconsider that data point; if possible the measurement should be duplicated. If this is not possible (for instance, it is 3:00 a.m. and your lab report is due that morning), you can use the error bars to support elimination of that point from the data pool. A new best-fit line for the data, calculated without the erroneous data point, is shown in Figure 10-6(b). Note that the slope and intercept for this new line is significantly different from that for the original data. With only a limited number of data points, it is advisable to remake measurements that have large deviations from the best-fit line. However, if there are a large number of points, then throwing out one or two has little effect on the overall quality of the data.

Chapter 10 Summary

- Common graphing methods include **bar charts**, **pie charts**, and **scatter plots**.
- In plotting data, the horizontal axis (*x*-**axis**) is used for the **independent variable** and the vertical *y*-**axis** (positioned perpendicular to the *x*-axis) is used for the **dependent variable**. Points have **(*x*, *y*) coordinates.**
- The slope is the "rise over run,"

$$m = \frac{\text{rise}}{\text{run}} = \frac{y_2 - y_1}{x_2 - x_1} = \frac{\Delta y}{\Delta x}.$$

- The **slope-intercept** equation for a line has the form $y = mx + b$, where m is the **slope** of the line and b is the *y*-**intercept**. The *y*-intercept is the value of *y* when $x = 0$ (where the line crosses the *y*-axis).
- The **point-slope** form of a line has the form $y - y_1 = m(x - x_1)$, where m is the **slope** and (x_1, y_1) is any **point** on the line.
- The equation of a line can be determined from *any* two points on the line: Points that lie between data points are determined by **interpolation**, and those that lie outside of the data range are determined by **extrapolation**.
- Relations of interest can be converted from their original form to a linear format so that measurements can be plotted as (*x*, *y*) pairs. Taking a root or logarithm of an equation is a common method of conversion.
- The best-fit linear relation to data can be accomplished with the Least Squares Method.
- Uncertainty in measurements is expressed graphically as an **error bar**.
- A graphing calculator can draw a scatter plot and find an equation of best fit (regression equation).

Practice Exercises Chapter 10 ———————————————

1. Plot this data using methods in the Chapter 10 Summary.

    a.

x	y
0	0
1	2
2	4
3	5
4	6
5	6.5
6	7
7	7.5
8	8
9	8
10	8
15	8

    b.

x	y
1	2.4
2	3.0
4	4.1
5	4.5
6	5.0
8	6.1
10	6.9

2. Plot the lines with these equations:

    a. $y = 2x + 1$

    b. $x + 2y + 2 = 0$

3. Determine the slope and y-intercept for the lines in exercises 1(b) and 2(b). Write the equation for the line in exercise 1(b) (you must estimate the best line for 1(b) by eye).

4. Using the graphs obtained from exercises 1 and 2, determine a value for y at the following values of x:

    a. $x = 3.0$ for line in 1(b)

    b. $x = 14.0$ for line in 1(b)

    c. $x = -4.0$ for line in 2(a)

    d. $x = 0.23$ for line in 2(a)

    e. $x = -2.0$ for line in 2(b)

    f. $x = 22$ for line in 2(b)

5. Determine the Least Squares best-fit line to the data in exercise 1(b). Compare this equation to the one you drew by eye in exercise 3.

6. Use a graphing calculator to draw a scatter plot for the given data. Find a regression equation using your calculator.

Time (s)	Pressure (torr)
0	24.0
1	18.1
2	13.7
3	10.3
4	7.8
5	5.9
6	4.5
7	3.4
8	2.6
9	1.9
10	1.5

7. Write an equation of a line with:

   a. $m = 5$ and $b = 4$.

   b. a slope of $\frac{2}{3}$ and a y-intercept of –2.

8. Write an equation of a line that

   a. passes through (1 , 5) and has a slope of $\frac{1}{3}$.

   b. passes through (–1 , 3) and (–2 , 5).

9. Use a graphing calculator to draw a scatter plot for exercise 1(b). Find a regression equation using your calculator. Does this agree with your answer to exercise 3 that you did by eye?

CHAPTER 11

Geometry and Trigonometry

"The sum of the square roots of any two sides of an isosceles triangle is equal to the square root of the remaining side!"
— The Scarecrow in *The Wizard of Oz*

Apparently the excitement of getting a new brain was too much for the Scarecrow. He mixed up his trigonometry and made an incorrect statement about isosceles triangles (triangles with two equal sides). He was on the right track, though. Perhaps he meant to say "The sum of the squares of the sides of a *right* isosceles triangle is equal to the square of the remaining side." Oh well, he did pretty well for someone with a brain made of straw.

Trigonometry involves relationships between the angles and sides of triangles. It may not be immediately obvious, but triangles are related to circles. Learning this connection will help you remember and use trigonometric functions. This skill will be valuable for solving problems in science that involve geometry and trigonometry.

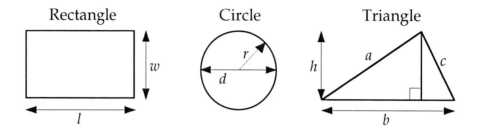

Figure 11-1 Rectangle: Perimeter, $P = 2l + 2w$; Area, $A = lw$. Circle: Perimeter, $P = \pi d$, Area, $A = \pi r^2$. Triangle: Perimeter, $P = a + b + c$, Area, $A = \frac{1}{2}bh$.

BASIC GEOMETRY

Most of the geometry that you will use in first courses of science will involve simple structures like rectangles, circles, and triangles. It is important that you learn the relations that govern boundaries of these structures. You are expected to know the basic characteristics of regular shapes, including **perimeter**, **area**, and **volume**.

The **perimeter** is the distance around the outside of a closed shape. For straight-sided shapes, you must sum the lengths of the sides. A rectangle has a length (l) and a width (w). The perimeter is simply $P = 2l + 2w$. For a triangle, you sum the lengths of the three sides to determine the perimeter. A circle's perimeter is called the **circumference**, C. The circumference of a circle is equal to the **diameter** (d, the distance measured through the center of the circle) times π. Since the diameter is twice the **radius**, this relation becomes $C = \pi d = 2\pi r$. Refer to Figure 11-1 for a summary of simple geometry of rectangles, circles, and triangles.

The **area** (with units of length^2) of a regular shape is how much space the shape occupies in two dimensions. The area of a circle is $A = \pi r^2$. The area of a rectangle is just $A = lw$. Since a right triangle is half the area of a rectangle (you make a rectangle by putting two right triangles together), the area of a triangle must be $A = \frac{1}{2}lw$. For any triangle, we call l and w the **height** and the **base**. The area of a triangle is $A = \frac{1}{2}bh$. Just remember that the base and height are always perpendicular to each other in a triangle. For triangles with a 90° angle, select a leg as the base and the perpendicular leg as the height. If the triangle does not have a 90° angle, treat it like the one in Figure 11-1.

The previous discussion applied to two-dimensional regular shapes. In three dimensions, shapes occupy **volume** (with units of length³). Consider a rectangle that you transform into a box (a **rectangular prism**); the box has length, width, and height. The box volume is simply the product of these factors, $V = lwh$. (Sometimes height is expressed as depth, $V = lwd$.) A **triangular prism** has triangles as the bases (just as the rectangular prism has rectangles as the bases), so the volume is the area of the base times the depth of the prism, $V = \left(\frac{1}{2}bh\right) \times d$. Depth is used here instead of height to reduce confusion with the height of the triangle used to calculate its area.

In general, the formula for the volume of a prism is the area of the base times the height, where B stands for the area of the base:

$$V_{prism} = h(\text{area of the base}) = hB$$

A cylinder has two circular bases, like a prism. The formula for volume is the same as that for the prism in the previous example, except the area of the base B is πr^2,

$$V_{cylinder} = hB = h\pi r^2$$

Another common solid object is the sphere. The volume of a sphere is determined only by the radius:

$$V_{sphere} = \frac{4}{3}\pi r^3$$

See Figure 11-2 for a summary of volumes for common solid objects. Note in all cases that the units of volume involve a length cubed (like cm³ or in³) and that area has units of length squared (cm², in², etc.).

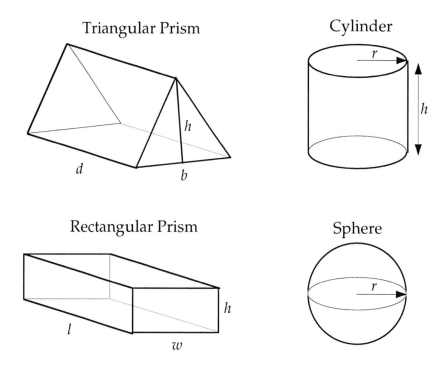

Figure 11-2 Volumes for solid objects. Rectangular Prism: $V = lwh$; Triangular Prism: $V = \left(\frac{1}{2}bh\right) \times d$; Cylinder: $V = h\pi r^2$; Sphere: $V = \frac{4}{3}\pi r^3$.

Right Triangles

Right triangles are very important components in trigonometry. A right triangle is simply a triangle with one **right angle**, an angle measuring 90°. The side opposite the right angle is called the **hypotenuse** (c) and the other two sides are called **legs** (a and b). The **Pythagorean Theorem** provides a relationship between the values of the sides of a right triangle. If you know the value for two sides of a right triangle, you can find the length of the remaining side using this relation. Consistent with the variables used above, the Pythagorean Theorem becomes:

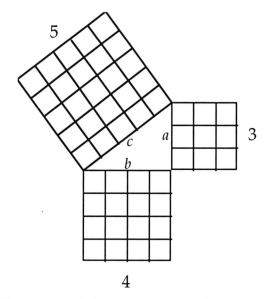

Figure 11-3 An illustration of the Pythagorean Theorem. The square of the hypotenuse (25 blocks) is equal to the sum of the squares of the other sides (16 plus 9 blocks).

$$a^2 + b^2 = c^2$$

The square of the hypotenuse length is equal to the sum of the squares of the remaining sides. Sounds a lot like what the Scarecrow was trying to tell the Wizard, doesn't it?

The Pythagorean Theorem is demonstrated in Figure 11-3, where the right triangle has legs with lengths 3 and 4 and the hypotenuse has length 5. You can see from the figure that $(3)^2 + (4)^2 = 9 + 16 = 25 = (5)^2$, the square of the hypotenuse length. The hypotenuse is the side opposite the right angle and is always substituted for c, and the legs of the right triangle are always substituted for a and b.

Important consequences of the Pythagorean Theorem are special right triangles with angles 45°-45°-90° and 30°-60°-90° (Figure 11-4). These triangles play an important part in the circle trigonometry discussed later in

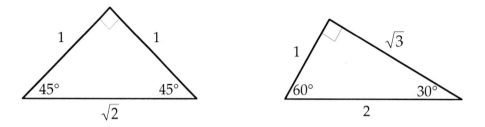

Figure 11-4 Left: The 45°-45°-90° triangle has side ratios $1:1:\sqrt{2}$. Right: The 30°-60°-90° triangle has side ratios $1:\sqrt{3}:2$.

this chapter. In a 45°-45°-90° right triangle, the hypotenuse is always $\sqrt{2}$ times the leg length. This right triangle is **isosceles**, which means it has two legs of the same length opposite two angles with the same value (45°). The ratio of the sides in an isosceles right triangle is $1:1:\sqrt{2}$. In a 45°-45°-90° triangle with a 5 cm leg, the hypotenuse would be $5\sqrt{2}$ cm. This can be verified with the Pythagorean Theorem:

$$5^2 + 5^2 = \left(5\sqrt{2}\right)^2 = 50$$

Similarly, in a 45°-45°-90° triangle with a 12 mm hypotenuse, the legs would each measure $\frac{12}{\sqrt{2}}$ (divide the hypotenuse by $\sqrt{2}$), or $6\sqrt{2}$ mm. When finding sides of a 45°-45°-90° triangle, you simply multiply or divide by $\sqrt{2}$.

In a 30°-60°-90° triangle, the shortest side (short leg) is always opposite the 30° angle, and the longer leg (but not the hypotenuse) is always opposite the 60° angle. The ratio of the sides is $1:\sqrt{3}:2$. In other words, the hypotenuse is always twice the length of the short leg; to find the long leg (side opposite the 60° angle), multiply the short leg by $\sqrt{3}$. If you have a 30°-60°-90° triangle with a 4 cm hypotenuse, the short leg would measure 2 cm and the long leg would measure $2\sqrt{3}$ cm. Again, this can be verified using the Pythagorean Theorem:

$$2^2 + \left(2\sqrt{3}\right)^2 = 4 + 12 = 4^2$$

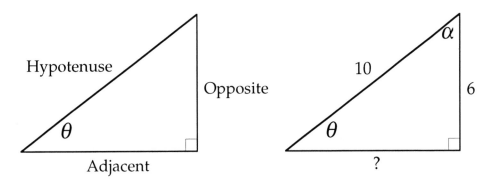

Figure 11-5   Right angles (see text for description of labels).

Right Triangle Trigonometry

Right triangle trigonometry involves the ratio of the sides in a right triangle. The right triangle in Figure 11-5 has labels for the *hypotenuse* (H), the *opposite side* (O), and the *adjacent side* (A). These labels are chosen as they relate to the angle θ (theta): O is the side opposite the angle θ, and A is the side adjacent to θ (but not the hypotenuse). The three main trigonometric functions **sine** (sin), **cosine** (cos), and **tangent** (tan) are defined using these labels. A short phrase may help you remember the relations between these functions: "**O**scar **h**ad **a** **h**eck **o**f **a** time **c**atching **s**almon." The first letters of each word represent **o**pposite, **h**ypotenuse, **a**djacent, **t**angent, **c**osine, and **s**ine in these equations:

$$\frac{\text{opposite}}{\text{hypotenuse}} = \frac{O}{H} = \sin\theta \qquad\qquad \frac{\text{Oscar}}{\text{had}} = \text{salmon}$$

$$\frac{\text{adjacent}}{\text{hypotenuse}} = \frac{A}{H} = \cos\theta \qquad\qquad \frac{\text{a}}{\text{heck}} = \text{catching}$$

$$\frac{\text{opposite}}{\text{adjacent}} = \frac{O}{A} = \tan\theta \qquad\qquad \frac{\text{of}}{\text{a}} = \text{time}$$

Read the sentence starting with Oscar and ending with salmon (go down the left side and up the right side). Stupid but effective.

Note that the tangent can be determined as the ratio of $\sin\theta$ to $\cos\theta$,

$$\tan\theta = \frac{O}{A} = \frac{\sin\theta}{\cos\theta}$$

There are three more trigonometric functions based on sine, cosine, and tangent: their reciprocals **cosecant** (csc), **secant** (sec), and **cotangent** (cot). Sorry, we don't have a silly phrase to help you remember the relations between these functions:

$$\csc\theta = \frac{\text{hypotenuse}}{\text{opposite}} = \frac{H}{O}$$

$$\sec\theta = \frac{\text{hypotenuse}}{\text{adjacent}} = \frac{H}{A}$$

$$\cot\theta = \frac{\text{adjacent}}{\text{opposite}} = \frac{A}{O}$$

Refer to the triangle in Figure 11-5 with the unknown side. You should be able to write values of the six trigonometric functions for both α and θ. Let's find the values of the six trigonometric functions for α. Using the Pythagorean Theorem to find the measure of the third side, we find that the missing leg of the right triangle is 8 cm. Therefore:

$$\sin\alpha = \frac{\text{opposite}}{\text{hypotenuse}} = \frac{O}{H} = \frac{8}{10} = \frac{4}{5} \qquad \csc\alpha = \frac{\text{hypotenuse}}{\text{opposite}} = \frac{H}{O} = \frac{10}{8} = \frac{5}{4}$$

$$\cos\alpha = \frac{\text{adjacent}}{\text{hypotenuse}} = \frac{A}{H} = \frac{6}{10} = \frac{3}{5} \qquad \sec\alpha = \frac{\text{hypotenuse}}{\text{adjacent}} = \frac{H}{A} = \frac{10}{6} = \frac{5}{3}$$

$$\tan\alpha = \frac{\text{opposite}}{\text{adjacent}} = \frac{O}{A} = \frac{8}{6} = \frac{4}{3} \qquad \cot\alpha = \frac{\text{adjacent}}{\text{opposite}} = \frac{A}{O} = \frac{6}{8} = \frac{3}{4}$$

Note that the leg ratios correspond to the triangle shown in Figure 11-3. Now you have six expressions you can use to determine the value of the angle α. Let's use the $\sin^{-1}$ function to find α (the calculator key located above the [sin] key).   Use this key only when **looking for an angle**.   The notation sin⁻¹ means the *inverse* trigonometric function of sine, not the reciprocal! To find the reciprocal of the sine function, use cosecant:

$$\sin\alpha = \frac{4}{5}$$

$$\alpha = \sin^{-1}\left(\frac{4}{5}\right) = 53.13°$$

```
Normal Sci Eng        sin⁻¹(4/5)
Float 0123456789          53.13010235
Radian Degree
Func Par Pol Seq
Connected Dot
Sequential Simul
Real a+bi re^θi
Full Horiz G-T
```

BASIC CIRCLE TRIGONOMETRY

Consider a circle with its center at the coordinates (0, 0) and a **radius** of exactly **1 unit.** Imagine that you make the circle by rotating an arrow that has a length of **1** and lies initially on the *x*-axis, so that the point will rest at the coordinates (1, 0). If you rotate the arrow through one complete revolution, the point travels the entire circumference of the circle, to return back to the starting position. This circle with a radius of 1 is called the **unit circle**, and it is shown in Figure 11-6.

The angle made between the positive *x*-axis (initial side) and the arrow (terminal side) is often designated with the Greek letter θ (theta). This angle can be expressed in two ways, using either **degrees** or **radians**. The **degree** as a unit of angular measure came from ancient mathematicians who divided the circle into 360 equal parts. So one complete revolution around a circle

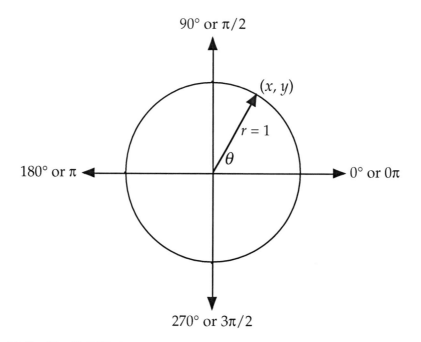

Figure 11-6   The Unit Circle.

represents 360°. A more natural unit of angular measure is generated by wrapping a number line around the unit circle, starting at the positive x-axis. Since the circumference of the unit circle is 2π ($C = 2\pi r$, $r = 1$), the wrapped number line divides the circle into 2π, or a bit more than 6 parts. In other words, if you travel one unit along a unit circle, beginning at the positive x-axis, the arc length is one unit (distance along the edge of the circle) and the angle measures one **radian.**

An angle that moves in a counterclockwise direction from the positive x-axis is a **positive angle**, while a clockwise rotation from the positive x-axis is a **negative angle.** In addition, **coterminal** angles are angles with the same terminal side. You can find coterminal angles by adding or subtracting multiples of 2π or 360°. For instance, 120°, 480°, and −240° are all coterminal angles because they have the same terminal side.

To convert angles between degrees and radians, recall that π radians are equal to half of the unit circle, 180°:

$$\theta(\text{degrees}) \times \frac{\pi \text{ radians}}{180°} = \theta(\text{radians})$$

$$\theta(\text{radians}) \times \frac{180°}{\pi \text{ radians}} = \theta(\text{degrees})$$

For example, to convert $\theta = 90°$ to radians:

$$\theta = 90° \times \frac{\pi \text{ radians}}{180°} = \frac{90°}{180°} \times \pi \text{ radians} = \left(\frac{1}{2}\right) \times \pi \text{ radians} = \frac{\pi}{2} \text{ radians}$$

To change $\theta = \dfrac{2\pi}{3}$ to degrees:

$$\theta = \frac{2\pi}{3} \times \frac{180°}{\pi} = \frac{360°}{3} = 120°$$

Note that angles missing the degree symbol are assumed to be in radians (like $\frac{2\pi}{3}$).

Every point along the unit circle has coordinates, or (x, y) values (refer to Figure 11-6). These values are related to the angle θ through the trigonometric functions cosine and sine. Consider the unit circle to have a radius r (equal to 1 in this case). The ratios of x and y to r depend on θ:

$$\cos\theta = \frac{x}{r} \quad \text{and} \quad \sin\theta = \frac{y}{r}$$

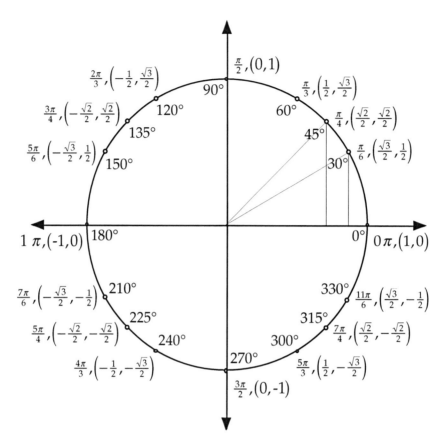

Figure 11-7 Common angles with radian equivalents and positions on the unit circle.

Since $r = 1$ for the unit circle, the x-coordinate becomes $\cos\theta$, and the y-coordinate becomes $\sin\theta$. Some of the common points along the unit circle are derived simply from the 30°-60°-90° and 45°-45°-90° triangles discussed on page 102.

Look at 30° or $\frac{\pi}{6}$ on the unit circle shown in Figure 11-7. If you drop a perpendicular line to the x-axis, you have formed a 30°-60°-90° triangle with a hypotenuse of 1, since the radius of the circle is 1. The short leg would measure $\frac{1}{2}$(half the hypotenuse) and the long leg would measure $\frac{\sqrt{3}}{2}$ (the short leg times $\sqrt{3}$). The (x, y) value at $\frac{\pi}{6}$ would be $\left(\frac{\sqrt{3}}{2}, \frac{1}{2}\right)$:

$$\cos 30° = \cos\frac{\pi}{6} = \frac{\sqrt{3}}{2} \quad \text{and} \quad \sin 30° = \sin\frac{\pi}{6} = \frac{1}{2}$$

Similarly, look at 45° or $\frac{\pi}{4}$ on the unit circle. If you drop a perpendicular line to the x-axis, you have formed a 45°-45°-90° triangle with a hypotenuse of 1 (the radius of the circle is 1). The legs would measure $\frac{1}{\sqrt{2}} = \frac{\sqrt{2}}{2}$ (divide hypotenuse by $\sqrt{2}$). The ordered pair at $\frac{\pi}{4}$ would be $\left(\frac{\sqrt{2}}{2}, \frac{\sqrt{2}}{2}\right)$:

$$\cos 45° = \cos\frac{\pi}{4} = \frac{\sqrt{2}}{2} \quad \text{and} \quad \sin 45° = \sin\frac{\pi}{4} = \frac{\sqrt{2}}{2}$$

It helps to learn a number of points along the unit circle that correspond to common angles ($\frac{\pi}{4}$, $\frac{\pi}{2}$, π, $\frac{3\pi}{2}$, etc.).

Looking at the unit circle, you should notice that sine is *positive* in quadrants I and II and *negative* in the others. Sine is positive when the y-coordinate is positive. Cosine is *positive* in the I and IV quadrants and *negative* in the others. So cosine is positive when the x-coordinate is positive. Tangent is *positive* in quadrants I and III. Can you see why?

$$\tan\theta = \frac{y}{x} = \frac{\sin\theta}{\cos\theta}$$

Sine and cosine are both positive in quadrant I and are both negative in quadrant III.

Trigonometric Identities

A trigonometric **identity** is a relationship that is always true for any given angle. Some common identities include the reciprocal identities of the trigonometric functions:

$$\sec\theta = \frac{1}{\cos\theta} \qquad \csc\theta = \frac{1}{\sin\theta} \qquad \cot\theta = \frac{1}{\tan\theta}$$

$$\tan\theta = \frac{\sin\theta}{\cos\theta} \qquad \cot\theta = \frac{\cos\theta}{\sin\theta}$$

For example,

$$\sec\left(\tfrac{\pi}{3}\right) = \frac{1}{\cos\left(\tfrac{\pi}{3}\right)} = \frac{1}{\left(\tfrac{1}{2}\right)} = 2$$

This can be typed into your graphing calculator directly or by using the reciprocal key, {x^{-1}}. Make sure the calculator is set for Radian mode. Do **not** use the {**cos**$^{-1}$} key! That key is only used to find an angle. In this case, you are looking for the value of the secant at an angle of $\tfrac{\pi}{3}$.

An important consequence of the Pythagorean Theorem is a trigonometric identity that relates sine to cosine. If you apply the Pythagorean Theorem to the right triangle in our unit circle (see Figure 11-6), you find that $1 = x^2 + y^2$, so that

$$1 = (\cos\theta)^2 + (\sin\theta)^2$$

or

$$1 = \cos^2\theta + \sin^2\theta$$

This relationship is the **Pythagorean Identity,** and it is true for any angle that you substitute for θ. Note that $\sin^2\theta$ is defined as $(\sin\theta)^2$, and $\cos^2\theta$ is the same as $(\cos\theta)^2$. The Pythagorean Identity allows a calculation of $\sin\theta$ if you know $\cos\theta$, and *vice versa*. For example, if you know $\cos(\theta) = \frac{4}{5}$, then you can determine the sine using the Identity:

$$\sin^2\theta = 1 - \cos^2\theta = 1 - \left(\tfrac{4}{5}\right)^2$$
$$\sin^2\theta = 1 - \tfrac{16}{25} = \tfrac{9}{25}$$
$$\sin\theta = \pm\,\tfrac{3}{5}$$

The sign will depend on the quadrant of the unit circle; sine is positive in the I and II quadrants.

Graphs of Cosine and Sine

Graphs of the functions $y = \cos x$ and $y = \sin x$ are easy to generate using your graphing calculator. The angle in radians is x, and the value from the unit circle is y:

$$y = \cos x$$

Input the function $\cos(x)$ under {**Y =**} then press {**GRAPH**}. You can generate values of y for various values of x entered under the X column in {**2nd TABLE**}. For example, values of $x = 0$, $\frac{\pi}{2}$, π, $\frac{3\pi}{2}$ yield $y = 1$, 0, –1, 0 as expected from inspection of the unit circle. If you wish to see these points on your graph, input the ordered x, y pairs (0, 1), $\left(\frac{\pi}{2}, 0\right)$, $(\pi, -1)$, $\left(\frac{3\pi}{2}, 0\right)$ under L_1 and L_2 after entering {**STAT 1:Edit…**}. Make sure Plot1 is turned ON in

{2nd **STATPLOT**}. We've chosen a convenient window that is [Xmin = $-\frac{\pi}{2}$, Xmax = $\frac{5\pi}{2}$]×[Ymin = –2, Ymax = 2].

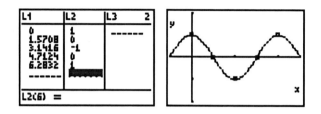

Follow the same procedure to view the sine function,

$$y = \sin x$$

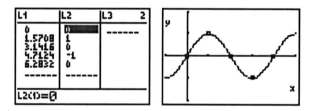

The graphs of $y = \cos(x)$ and $y = \sin(x)$ are **periodic functions**, which means the cycle will repeat over regular intervals. In this case, the graph will repeat itself every 2π radians or 360°. The **domain** is what goes into the function. You can substitute any real number into the equation for your angle x in radians or degrees.

$$\text{Domain} = \left\{x \mid x \in \text{Real Number System}\right\}$$

The symbol $\in$ means "is an element of." The **range** represents what comes out of the function. This is the y value:

$$\text{Range} = \left\{y \mid -1 \le y \le 1\right\}$$

For values of y along the unit circle, the range would include numbers between, and including, –1 and 1.

When using these functions, it is important to be aware of how your calculator treats numbers. Angles can be expressed in either radians or degrees, and it is critical that you set your calculator to operate in the mode that matches your input. This is usually changed using a mode key or key sequence. **Check your manual** to make sure you know how to operate your calculator in the proper mode. Some calculations will be easier to perform in degrees, while some formulas will require radians. Learn the difference and practice using your calculator in both modes.

VECTOR BASICS

Look again at the arrow shown in Figure 11-6. It has a specific length (1 in this case) and **direction**. The direction is indicated by the angle θ relative to the positive x-axis. If you imagine that the unit circle is a compass with $0°$ corresponding to the direction of east, the arrow points generally toward the northeast (north is at $90°$). These traits of length and direction are characteristics of vectors. A **vector** has a numerical value (with units) and a direction, such as 55 mph due west. Examples of vectors include electric and magnetic fields, forces like gravity, dipole moment, acceleration, and velocity. A **scalar** has only a numerical value and unit, like $5, 3 inches, or 250 pounds.

One way to indicate a vector is to place an arrow over a variable, such as $\vec{A}$. The vector $\vec{A}$ has a **magnitude** $|A|$; the bars indicate to take the absolute (non-negative) value of the vector length. Figure 11-8(a) shows a vector $\vec{A}$ as it relates to x- and y-axes. By dropping lines from the arrow tip perpendicular to the axes you perform "projections" of the vector onto the axes. The lengths of the projections are related to the vector by the Pythagorean Theorem:

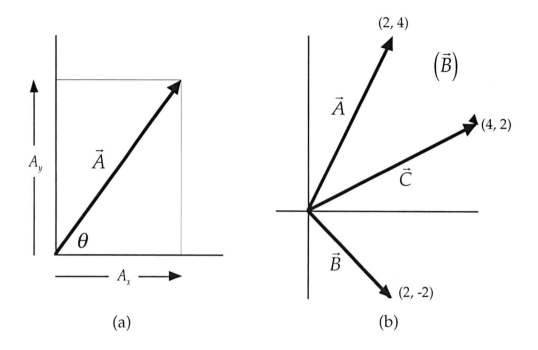

(a)   (b)

Figure 11-8 Characteristics of vectors; (a) projections of vector $\vec{A}$, (b) addition of vectors $\vec{A}$ and $\vec{B}$.

$$A_x = |A|\cos\theta$$

$$A_y = |A|\sin\theta$$

$$|A| = \sqrt{A_x^2 + A_y^2}$$

$$\theta = \tan^{-1}\left(\frac{A_y}{A_x}\right)$$

Scalars and vectors can be multiplied together to change the length and possibly the direction of the vector. For the product $s \times \vec{A}$, if the scalar s is positive, then the resultant vector has length $s|A|$ and points in the *same* direction as $\vec{A}$. If s is negative, then the resultant vector has length $s|A|$ but points in the *opposite* direction $(-\vec{A})$. For example,

$$-2 \times (30 \text{ mph east}) = -(60 \text{ mph east}) = 60 \text{ mph west}$$

The order does not matter for addition of vectors; they are said to **commute**:

$$\vec{A} + \vec{B} = \vec{B} + \vec{A}$$

It also does not matter how additions are grouped; vectors are **associative**:

$$\left(\vec{A} + \vec{B}\right) + \vec{C} = \vec{A} + \left(\vec{B} + \vec{C}\right)$$

To subtract vector $\vec{B}$ from $\vec{A}$, reverse the direction of $\vec{B}$ and add to $\vec{A}$:

$$\vec{A} - \vec{B} = \vec{A} + \left(-\vec{B}\right)$$

Notice that when vectors of equal magnitude and opposite direction are added, the resultant is equal to zero:

$$\vec{A} + \left(-\vec{A}\right) = \vec{A} - \vec{A} = 0$$

When adding vectors, the tip of one vector is placed at the tail of the other (see Figure 11-8b). The vector $\vec{A}$ can be represented by the coordinates of the tip (2, 4). Vector $\vec{B}$ has tip coordinates (2, –2), and its tail is placed at the tip of vector $\vec{A}$. The sum $\vec{A} + \vec{B}$ results in $\vec{C}$, found by adding the x values and the y values:

$$\vec{C} = \vec{A} + \vec{B} = (2,4) + (2,-2) = (2+2, 4-2) = (4,2)$$

The other way to look at this is with the projections:

$$\vec{A} : A_x = 2, \ A_y = 4$$

$$\vec{B} : B_x = 2, \ B_y = -2$$

$$\vec{C} : A_x + B_x, \ A_y + B_y = 2+2, \ 4-2 = (4,2)$$

Therefore, $C_x = 4$, $C_y = 2$, and $|C| = \sqrt{4^2 + 2^2} = \sqrt{20}$.

Chapter 11 Summary

- Trigonometric functions are related to circles and triangles.
- Characteristics of regular shapes include **perimeters**, **areas**, and **volumes**.
- A **right triangle** has a hypotenuse opposite the right angle (90°).
- The **Pythagorean Theorem** relates the square of the hypotenuse of a right triangle to the squares of the remaining two sides.
- Be careful to set your calculator to operate in the correct mode for either degrees or radians.
- A circle has a radius and contains 360° (2π radians). The unit circle is used to generate the trigonometric functions **sine** and **cosine**. The ratios of x and y to the radius depend on the angle θ through the functions sine and cosine. The angle 0° corresponds to the point $(1, 0) = (\cos 0°, \sin 0°)$ on the unit circle.
- For a circle with a radius, r, other than 1, coordinates along the circle become $(x, y) = (r\cos\theta, r\sin\theta)$.
- From the Pythagorean Theorem, the identity $1 = \cos^2\theta + \sin^2\theta$ can be derived.

- **Vectors** have **magnitude** and **direction**. A vector $\vec{A}$ that makes angle θ with the x-axis has magnitude $|A| = \sqrt{A_x^2 + A_y^2}$, where $A_x = |A|\cos\theta$ and $A_y = |A|\sin\theta$. The angle θ is found from $\theta = \tan^{-1}\left(\dfrac{A_y}{A_x}\right)$. **Scalars** have magnitude but no direction.
- To add vectors, sum the x and y projections: $(C_x, C_y) = (A_x + B_x, A_y + B_y)$ for $\vec{C} = \vec{A} + \vec{B}$.

Practice Exercises Chapter 11 ————————————————

1. Convert these angles in **radians** to angles in **degrees**:

   a. π

   b. $\frac{\pi}{4}$

   c. $-\frac{3\pi}{2}$

   d. 2.43

2. Convert these angles in **degrees** to angles in **radians**:

   a. 135°

   b. −270°

   c. 30°

   d. 257.3°

3. Compute these trigonometric functions (use your calculator only when necessary):

   a. $\sin(2\pi)$

   b. $\cos\left(-\frac{\pi}{2}\right)$

   c. $\sin(45°)$

   d. $\tan(60°)$

4. Use your skill with trig functions to determine the following for a right triangle:
   a. For a 23° angle, the adjacent side = 2.5. How long is the hypotenuse?
   b. For an unknown angle, the adjacent side = 3.0 and the hypotenuse is 4.0. What is the angle?
   c. For an unknown angle, the adjacent side = 2.2 and the opposite side = 4.4. How long is the hypotenuse?

    d. The triangle has a 62° angle and a hypotenuse = 5.0. What are the lengths of the other two legs of the triangle? What are the values of the other two angles?

5. a. A vector originates at $(0, 0)$ and terminates at $(-4, 3)$. What is the length of the vector, and what angle is made between the vector and the x-axis?

    b. Add the vectors $\vec{A}$ and $\vec{B}$ if $\vec{A} = (-2, 2)$ and $\vec{B} = (4, 1)$.

    c. Determine $\vec{A} - \vec{B}$ for the vectors in exercise 5(b).

6. Determine the perimeter, area, or volume where appropriate:

    a. Calculate the perimeter and area for a square with a side = 4.2 cm.

    b. Calculate the area of a circle with a diameter of 6.0 in.

    c. Calculate the volume of a box with dimensions $2 \times 4 \times 5$.

    d. Calculate the volume of a cylinder with radius = 2.0 and height = 10.0.

    e. Calculate the volume of a spherical droplet of water with radius of 1.0 micrometer.

    f. Calculate the area of a triangle with base length 10.0 in and height 2.00 in.

7. Find the measurements of the other sides of the triangle described:

    a. a 30°-60°-90° triangle with hypotenuse 6 inches.

    b. a 30°-60°-90° triangle with long leg $6\sqrt{3}$ cm.

    c. a 45°-45°-90° triangle with hypotenuse 8 ft.

    d. a 45°-45°-90° triangle with leg $2\sqrt{2}$ m.

8. Find the value of the six trigonometric functions if the terminal side passes through the point $(-4, 3)$. *Hint:* Drop a perpendicular to the x-axis to construct a right triangle.

9. Give the possible values for θ between 0 and 2π:

a. $\sin\theta = \dfrac{1}{2}$ c. $\cos\theta = -1$

b. $\tan\theta = 1$ d. $\cos\theta = \dfrac{\sqrt{2}}{2}$

10. Find the five other trigonometric functions for θ, given that θ is a first quadrant angle and $\tan\theta = \frac{4}{7}$. Can you do the same thing if θ is in the third quadrant?

CHAPTER 12

Solving Problems

In this chapter we will focus on some basic strategies that will help you approach problems with a good chance of success. If you approach solving problems from the standpoint that there is some secret formula to follow, then you are certainly doomed. You become a good problem solver only after much practice and hard work (sorry to burst your bubble). Solving problems quickly and correctly depends on your knowledge of the subject, your skill with mathematical techniques, and the insights that you gain by doing many and different problems. Research into problem solving indicates that the most successful problem solvers rewrite problems in a different, more familiar form. Often drawing a little picture is enough to get the expert problem solver close to the answer. Anyone can be a good problem solver! Start now to build your skills.

When faced with a problem to solve, the best piece of advice is to look before you leap. If you learn to proceed slowly with the confidence that comes from knowing your subject, you have a much better chance of determining the right answer. You must **read** the problem carefully (don't skim it or make any assumptions as you go). A common complaint of students after an exam is that they just didn't read the problems carefully enough to know what to do to get them right. Often students try to solve the problem without really understanding what information is given and what they need to determine the answer. Strangely, some students try to solve problems on science exams without writing *anything*. Don't fall into that trap.

Use this basic approach for any problem:

1. Read the problem word for word.
2. As you read, write down the information given in the problem. Some prefer to underline or circle words and symbols in the problem, but you might find it more useful to write down variables and what they are equal to.
3. It might be appropriate to draw a little picture that you can label with values for known quantities.
4. Write down explicitly what information is being sought. Most often this is one variable that will have a particular value. Write down the anticipated units for the answer.
5. Write down the relationship between the information you know and the desired answer. This will usually be an equation in a familiar form.
6. Rearrange the equation in a convenient form that allows you to solve for the desired parameter. Put the numbers you know into the formula. Doing this early will identify conversions that you need.
7. Always carry units in the calculation, and make appropriate unit conversions so that your answer has the proper units.
8. Check your answer to make sure it is reasonable (more on making estimates in Chapter 13).

Here is an example to try with the above approach in mind:

> How many feet does a swallow fly in 20 minutes at an average air speed of 30 miles per hour?

- Write down what you know: $t = 20$ min $\quad v = 30$ mph
  (It is tempting to write down "swallow," but since the type of bird has no real bearing on the problem, don't write it down—it will only waste time.)

- You might draw a little picture of a flying bird, with a vector labeled 30 mph. You could add a longer line that represents the total distance flown.
- Write down what you seek: distance, $d = ?$ ft
- Write down a relation that you know:

  speed = distance/time

  $$v = \frac{d}{t}$$

- Rearrange to solve for distance: $d = v \cdot t$

- Insert numbers: $d = \left(30\,\tfrac{mi}{hr}\right)(20\,min)$

- Change units with conversions: $d = \left(30\,\tfrac{mi}{hr}\right)\left(5280\,\tfrac{ft}{mi}\right)(20\,min)\left(\tfrac{1hr}{60min}\right)$

- Perform calculation: $d = 10\ mi\left(5280\,\tfrac{ft}{mi}\right) = 52{,}800$ ft

- Check your answer: Ten miles is a long way, but it seems reasonable. A million miles wouldn't be reasonable.

Note that in the example above information was always attached to a variable and the units were written down explicitly. The problem asked for the distance in feet, so it was necessary to convert the units of speed (velocity, v) from miles per hour to feet per hour. If the units had not been carried, then that important change might not have been made and the problem would have been answered incorrectly. Carry the units to avoid simple mistakes like that.

A subtle, yet very important, aspect of problem solving is being familiar enough with your subject to think about it in terms of the variables commonly used in calculations. If you try to cram before an exam, then you are less likely to get mathematical problems correct because you have not familiarized your mind with the equations and their variables. You must expose yourself to variables multiple times before you begin to think about scientific relations in terms of variables and equations. This is why it is better to study in smaller amounts every day than to try to learn most of the material right before an exam. Through constant exposure, you train yourself to recognize and value the information that appears in problems that you will solve. By doing exercises on you own (Ack! Homework!), you help develop the familiarity that you need to solve problems on exams.

If you present the preceding problem to a student with little exposure to distance/velocity problems, you will get interesting answers when you ask simple questions about the problem. If you ask "What does the problem ask for?" the student might answer "feet" instead of "distance." The student might write down "$s = 30$ mph" instead of using the variable for velocity, v. This confusion stems from a lack of exposure to the subject and lack of practice using the proper variables and equations. Even though this student probably knew the relationship between distance, time, and velocity from reading the textbook and hearing the lecture, it apparently didn't sink in. This student has not yet learned the material well enough to think about it and discuss it clearly. This student might get the problem right, but not by understanding the relationships involved. You can avoid this syndrome by learning variables and thinking in terms of variables and equations, which requires practice with the equations and numerous homework problems—but it is well worth the effort. You will be rewarded by your ability to read problems and understand right away what information you have and what information you seek.

Most problems require the use of one equation to solve them. With one equation, you can determine only one unknown quantity. For every unknown quantity you must have an equation—two equations for two unknowns, for instance. Advanced problems might require you to determine one quantity from one relation and then input that result into another equation to get the answer. Multistep problems will come easier to you after you familiarize yourself with the individual steps. The strategy of writing down what you know and what you seek will help you see quickly whether there are intermediate steps in the problem that must be solved. Purposefully expose yourself to these types of problems. They enhance your abilities by strengthening the connection of ideas through the use of equations.

Sometimes it will seem that you don't have the information that you need to solve a problem. This can happen when there are constants in the equations; the constants themselves are not given in the problem, so you have to look them up or remember them. It is a good idea to ask your instructor what constants you are required to know for an exam (commonly they are given in a table at the beginning or end of an exam).   Another instance when it appears that you need more information is when you have neglected a simple relation like the one between percentages: all percentages add to a total of 100% (also, all fractions must add to a total of 1).   You might forget this simple relation and get stuck while doing a problem.

When you hit a brick wall like this, you must have the experience behind you to guide your actions. A problem on an exam is never impossible, but it might be challenging. Maybe you have never put two equations together before, or you haven't done a percentage problem where one of the percentages was missing. The experience of homework (there's that word again) will help you out of those jams and allow you to think on your feet during an exam. Build those strengths now by doing many problems and watching the problem-solving techniques of others—your instructor, teaching assistants, tutors, or classmates. Look for tricks and shortcuts only *after* you have convinced yourself that you can do it the long way first. Once you really understand what you are doing, you will see the shortcuts while practicing problems on your own—homework, homework, homework!

Chapter 12 Summary

- Read problems carefully and write down what you know, what you seek, and what relation (equation) will get you there.
- Consider drawing a little picture and labeling it with values of variables that you know from the information given in the problem.
- Always carry units (sometimes the units of the desired result will suggest the equation that you need to use).
- Check your answer to make sure it is reasonable. If you have time, put the numbers through it twice to make sure you get the same answer.
- Familiarize yourself with variables and equations to build your ability to do problems (do your homework!).
- Every unknown needs an equation, and there will only be one unknown per equation—you must have a value for every other variable and constant.
- Practice, practice, practice.

Practice Exercises Chapter 12

1. A student finds a gold-colored, heavy rock having a mass of 22.0 g. The student fills a graduated cylinder with water to exactly 10.0 cm^3. When the rock is placed in the water, the new volume in the cylinder is 14.3 cm^3. The density of pure gold varies between 15.3 and 19.3 g/cm^3. Is this rock a piece of gold?

2. Laser light pulses from certain Nd:YAG lasers last only about 5 ns. If the speed of light is 2.998×10^8 m · s^{-1}, what is the length of a single laser pulse? If the pulse is shortened to 2 ps, how long is it?

3. Marbles with radii of 5.0 mm are glued side-by-side around the circumference of an earth-mover tire. The tire has a diameter of 10.0 ft. How many marbles encircle the tire?

4. The speed of sound in air is 1140 feet per second (at 25 °C). If lightning strikes exactly 5.0 miles from you, how long will it take to hear the thunder? Determine a general equation to help you determine the distance of lightning strikes from a measure of the time between flash and thunderclap.

5. In 1924 the French physicist Louis de Broglie made the somewhat startling suggestion that *all* moving particles, from electrons to city buses, have a wavelength λ. That wavelength is determined from the constant h (6.63×10^{-34} J · s), the particle mass m, and the velocity v:

$$\lambda = \frac{h}{mv}$$

What is the wavelength of a 150 g baseball thrown by a pitcher at 90 mph? (*Hint*: Remember that 1 J = 1 kg · m^2 · s^{-2}).

6. You want to wallpaper the 9 ft × 9 ft wall in your bedroom. The rolls of wallpaper are 18 in wide and 20 ft long. How many rolls do you need?

7. An ice skating rink is 200 ft × 50 ft, covered with a 1.0 in layer of ice. What is the total weight of ice in pounds if the density of the ice is 0.92 g/cm³?

8. Lemon Biscuits:

300 g flour	1 egg
180 g butter	1 pinch salt
70 g sugar	1 grated lemon rind
1 tsp vanilla	Bake 15-20 min. at 170-190 °C

   Convert this recipe to weights in ounces, assuming there are 28 g in 1 ounce. Convert the oven temperature to Fahrenheit.

9. Air pressure at sea level is approximately 14.7 psi (pounds per square inch). What is the maximum weight of argon gas in a column of air with a base of one square mile? The composition of air is 75.5% nitrogen and 23.1% oxygen.

10. In the surveyor's "chain system" of measurements, 1 link = 7.92 in, 1 chain = 100 links, and 1 furlong = 10 chains. The Kentucky Derby is a horse race of 10 furlongs. How many miles is this?

CHAPTER 13

Making Estimates

After you have mastered a basic understanding of how to solve problems, it is advantageous to learn how to make estimates. Estimates are useful in a number of situations. You should be able to estimate your answers to make sure that the number determined with your calculator is reasonable. You can usually spot calculator errors when the calculated result is orders of magnitude different than your estimate. Estimates can be useful on multiple choice tests where selections often differ by more than 10%. A quick estimate is sometimes faster than punching numbers into a calculator, and just as good if all you are trying to do is match a selection listed below the problem.

The key to good estimating is rounding numbers up or down to obtain values that are easier to multiply or divide. You have probably done this many times without thinking about applying it to scientific problems. For example, suppose you are planning for a party; you expect about 20 people over for chips, dips, and soft drinks. To estimate how much money the party will cost, assume that everyone will drink two drinks, for a total of 40, so seven six-packs will cover them. Five big bags of chips and three containers of dip will be enough (as long as your guests have already had dinner!). You saw drinks on sale for $1.89 a six-pack, so you estimate seven at $2 for a total of $14. Five bags of chips at $3 per bag (really $2.99) is $15, and the dip at $3 per container (really $2.89) adds another $9. The estimated sum is $14 + $15 + $9 = $38. The real cost is $36.85, so you overestimated by a bit. Clearly, the estimate took less time than the real calculation and gave a number that was close enough to be useful.

You can do the same thing with calculations used in solving problems. The most important thing to practice is quick estimates using numbers expressed in scientific notation. If you can't multiply or divide exponential numbers without your calculator (Chapter 4), then you won't be able to make good estimates. Estimate an answer to the following problem:

How far away is the sun if it takes sunlight 8.3 minutes to reach Earth?

What you know: $t = 8.3$ min $\quad v = c$ (speed of light) $\approx 3 \times 10^8$ m/s

What you seek: distance, $d = ?$ meters

Relation: $v = \dfrac{d}{t}$

Rearrange: $d = v \cdot t$

Estimate: $d = \left(3 \times 10^8 \, \frac{m}{s}\right)(8\,\text{min})\left(60 \frac{s}{\text{min}}\right) = (24 \times 60)\left(10^8\right)$ m

$= (144 \times 10)\left(10^8\right) = 1440 \times 10^8$ m

This answer is equal to 1440 $\times$ 10^5 km, or 144 $\times$ 10^6 km (144 million kilometers). If this had been a multiple choice test with the following selections,

A) 10 million km   D) 150 million km
B) 50 million km   E) 200 million km
C) 100 million km

which would you choose as the right answer? The estimate took less time than real calculator crunching, and yielded the right answer (D).

Here is another example with more gymnastic manipulations, an equilibrium calculation with numbers in exponential notation. What is K_C for the formation of hydrogen iodide, if the equilibrium expression is determined to be:

$$K_C = \frac{[HI]^2}{[H_2][I_2]} = \frac{\left(1.6482 \times 10^{-2}\right)^2}{\left(2.9070 \times 10^{-3}\right)\left(1.7069 \times 10^{-3}\right)}$$

The first order of business is to separate the exponents so that the power of the result can be determined:

$$K_C \cong \frac{(1.65)^2}{(2.91)(1.71)} \times \frac{\left(10^{-2}\right)^2}{\left(10^{-3}\right)\left(10^{-3}\right)}$$

Note that the numbers were reduced to three significant figures. Now the exponents can be reduced, and we should look for obvious factors in the numerator and denominator. Since 1.65 is close to 1.71, we will eliminate 1.65 once from the numerator and remove 1.71 from the denominator:

$$K_C \cong \frac{1.65}{2.91} \times \frac{10^{-4}}{10^{-6}} \cong \frac{1.65}{3} \times \left(10^2\right)$$

The denominator can be rounded to 3 for the division into 1.65. This is easy to do longhand, but you can estimate it by comparing some nearby "perfect" divisions. For example, $\frac{1.5}{3}$ is 0.5 and $\frac{1.8}{3}$ is 0.6, so the value of $\frac{1.65}{3}$ will be halfway between these results, namely, 0.55. Our estimate becomes:

$$K_C \cong 0.55\left(10^2\right) = 55$$

This was wildly successful, since the answer determined with five significant figures on a calculator is 54.748! Practice with these methods will cut down on the steps and time it takes to make manipulations.

When dealing with fractions, you can reduce their complexity by dividing the numerator and denominator by common factors. You need to see common factors quickly when making simplifications, and your speed depends on a thorough knowledge of the "times tables." If you rely on your calculator to do simple manipulations like 3×14, or $28 \div 7$, then you need to practice doing more calculations in your head.

Estimating roots and logarithms takes some additional practice. Taking complex roots can be difficult, but square roots are not too bad. For example, what is the square root of 70? Seventy is between the perfect squares 64 (8^2) and 81 (9^2), so the square root is between 8 and 9. Since 70 is less than halfway between 64 and 81, you might guess that the square root is less than 8.5, say about 8.3. That is a good guess since $\sqrt{70} = 8.37$ (no, we didn't cheat with a calculator, we really estimated 8.3).

You can use a similar approach with logarithms (base 10 logs; natural logs aren't so easy). What is log(779)? In scientific notation, 779 is written 7.79×10^2. Since 779 is between 100 (or 10^2) and 1000 (or 10^3), you know that log(779) is between 2 (log of 100) and 3 (log of 1000). The log scale is not linear, so interpolating between the extremes (2 and 3 in this example) is tricky. You will want to learn one more piece of information: the **log of 5.0 is 0.70** (log(2.0) = 0.30 and log(8.0) = 0.90 are also handy). So

$$\log(500) = \log(5.0 \times 10^2) = \log(5.0) + \log(10^2) = 0.70 + 2 = 2.70$$

Similarly,

$$\log(0.000050) = \log(5.0 \times 10^{-5}) = \log(5.0) + \log(10^{-5}) = 0.70 + (-5) = -4.30$$

Since 779 is between 500 and 1000, we estimate that log(779) is between 2.7 and 3. A good guess might be 2.85. That is a little low, since the correct answer is 2.89, but it was an excellent estimate (again, no calculator! Honest!).

Making useful estimates takes practice; if you have never tried making estimates, then it is time that you begin. The best time to try is when you are doing homework problems. Put your calculator aside and try to do the problems first by making estimates, then check your answer with the calculator. Also, follow along with examples in your text by estimating and marking numbers in the margin to match the numbers used by the author. After a few weeks of this, you will be a champion estimator, and you will strengthen your abilities to predict answers to problems without picking up your calculator. The ability to estimate is a valuable skill! Some of the insights that your instructors want you to gain in their courses involve making quick judgments and predictions. This type of thinking comes from understanding the material and knowing that the numbers look right. Good estimators are good students. Make both of these characteristics your goals.

Chapter 13 Summary

- Learn how to estimate your answers to make sure that the number determined with your calculator is reasonable and to develop a feel for the numbers.
- Make good rounding decisions to obtain values that are easier to multiply or divide in a complicated calculation.
- Good estimators multiply or divide exponential numbers quickly and without a calculator.
- Practice with these methods will cut down on the steps and time it takes to make estimates.
- Brush up on your "times tables" if you have forgotten them.
- Interpolating (guessing values between obvious extremes) in estimating roots and logarithms takes practice.
- The best time to practice is when you are doing homework problems. Start now and you will be rewarded.

Practice Exercises Chapter 13 ————————————————

1. Estimate the weight in tons (1 ton = 2000 lbs) of bauxite in a train car with approximate dimensions of 5 ft × 5 ft × 10 ft. Bauxite is an aluminum ore with a density of about 2.5 g/cm^3.

2. Estimate the energy of a 158-grain bullet traveling with a muzzle velocity of 2500 ft/s. Use the kinetic energy equation

$$E_{kinetic} = \tfrac{1}{2}mv^2$$

Express your answer in kJ (1J = 1 kg · m^2 · s^{-2}). One pound is equal to 7000 grains.

3. Vinegar is a solution of acetic acid with a hydrogen ion concentration, [H$^+$], near 0.004 mol/L. Estimate the pH of vinegar, using the definition

$$pH = -\log [H^+]$$

4. A 1-ft^3 plexiglas box is filled with chili beans. You measure 20 beans visible in a square inch of the box. Estimate the number of beans in the box to win a free trip to the East Texas Chili Festival.

5. Estimate your gasoline costs for a road trip to New Orleans for Mardi Gras, starting in Detroit, Michigan (a 1078-mile trip one-way). Your car gets about 28 mpg. Gas in your area sells for about $1.55 per gallon. If you maintain an average speed of 65 mph and only make four 20-minute pit stops, what is your total travel time (down and back)?

6. How may laps (down and back) in a 25-meter pool equals 1 mile of swimming?

Some Study Tips

How do you study? Does it work? Could you do different things to help you learn better? Your answers are probably something like: "1. Well, I really don't know. 2. Yeah, most of the time. 3. Sure, but what?" We don't pretend to have the answers for you, but we can offer some observations and suggestions that might help.

One of the biggest problems facing students and educators is that students have many different learning styles. What's worse, your instructors most probably have a *different* learning style than you do. That means that your instructor won't be making a special effort to present material in class that is customized to the way in which you might learn it best. The bottom line is this: Your performance in a class depends more on your personal relationship with the material than on the behavior of your instructor. It may be true that a good instructor can inspire, challenge, or interest you in a subject. These are positive attributes that help maintain your motivation in class, but they will not take the place of good studying and hard work.

So back to studying, what exactly is it? It is your interaction with the material to be learned in a class. If you have little interaction, you will most likely learn only a little. Even if you have a lot of interaction, you *are not guaranteed* to learn a lot. Often students "study" many hours before an exam, yet do poorly because the way they studied did not promote understanding of the material. Usually this means that they exposed themselves to individual tasks or skills but never added enough depth to their studying to allow them to actually learn the material. They knew the parts but could never see the whole.

In this chapter we will list for you some things related to studying that might be a help. Some of them will seem like common sense, but they are included here because you can never have too much common sense. The others will be ideas that have worked in the past for others, which you may adopt as you see fit. Remember, there is no secret or magic bullet when it comes to studying. It is likely that you will spend years developing study habits that work for you. Maybe some of the ideas presented here will help accelerate that process. Read on and try anything that looks like it might help.

Here comes the common sense part:

1. **Read your textbook**. Really read it, every word. Often students complain about exam questions, then look very sheepish when shown the exact problem from the text. Get in the habit of reading for **understanding**, not just exposure. Students who skim or "look over" the text are usually D students or worse. Take notes while you read—with a pencil in a separate notebook. Throw away the highlighter! Marking up your book with obnoxious day-glo inks will not help you learn better. More on this later.

2. **Go to class.** Not only will you see the material again, it will be through someone else's eyes. And that someone is very important. Your instructor interprets the course material and naturally stresses the most important points in class. Use this time to gather additional information about where you should emphasize your study of the material. Favorite examples or problems worked in class often will reappear on exams. Listen closely to your instructors because they are attempting to help you make connections about ideas in the course—to develop your depth of understanding. Don't go to sleep! If you find yourself slipping away, get more sleep the night before or concentrate harder on your notes. Try to predict what will be said next, or work on practice problems—do anything to stay awake.

3. **Take notes.** It will be very difficult to remember the wonderful and exciting things said by your instructors—unless you write them down.

Listen for voice inflections and write down phrases when your instructors seem to emphasize things. Copy notes carefully when derivations are done or equations are manipulated. If your instructor writes something on the board, you should, too. However, don't focus so much on writing that you miss what is said; develop a shorthand (don't attempt to take down things word for word) so you can write the important information while *listening* to the presentation of ideas. If this is hard for you, try taping the lectures and writing only things that are written on the board.

4. **Do your homework.** Sometimes it might not be clear what is expected of you with regard to homework. Most times homework is not collected for grading, nor are the problems covered in class. The fact of the matter is that you are expected to do homework, since it represents an active way for you to learn the material. If your instructor rarely assigns or talks about homework, don't make a mistake by assuming that it is not important. It is very important. You might find (no mystery here) that your instructor will often ask exam questions that are very similar to exercises in the text that appear as examples or homework.

5. **Stay ahead of your instructor.** Read your text before you attend class. That way the words used by your instructor won't be new, and the lectures will help you connect ideas that you have read about. When you read, write down things that are most difficult for you. Then when you attend your lecture, you have prepared your mind for the explanations. If the explanations do not materialize, then refer to suggestion 6. After class, look over your text again to emphasize new terms and relationships. Do this after every class.

6. **See your instructor.** If you read about an idea, hear that idea in class, look it over after class, and it still doesn't make sense to you—go see your instructor. Instructors set aside time to meet with their students (office hours). A few minutes with your instructor can often clear up little misunderstandings before they blossom into deep problems with the material. For some reason, students ignore this option and don't show up at their instructor's office until it is usually too late to reverse any damage done. Your instructor wants you to succeed, to see you understand the material. Take your list (see suggestion 5) to your

instructor for help. Your instructor can help you best when there are specific problems to address and work through.

7. **It makes a difference where you study.** If you are distracted while you study, then you will have a harder time retaining material. Don't watch television. Use quiet music (quiet, not necessarily boring) in the background if you need noise to keep you company. If friends dropping by or calling on the phone constantly interrupt you, then you will have to leave or turn off the phone. Find a place to spread out and get comfortable: a library, student lounge, or fast food restaurant (during off-hours when traffic is light). Above all make sure that the place you pick to study helps you concentrate on reading and working problems. It can be a chore to find such a place, but you will be rewarded by a deeper understanding of the material.

8. **Studying with others can help.** By talking over problems with other students in your class, you will use the new words and ideas that you are learning. This will help strengthen your understanding of the subject. It is important, however, that you study properly with others. Too often group study situations turn into opportunities for socializing. You must take steps to avoid this. One step is to pick a neutral site for studying that is not conducive to socializing (a cafeteria table during off-hours is a good place). Plan to go somewhere together after your study time to satisfy your urge to socialize and reward yourselves for sticking to the job of studying; be careful not to rush through the material in order to do the socializing.

It is best if you can structure your activities by deciding on specific tasks to accomplish between meetings. Preparing short practice exams (to be discussed more fully below) is one aid that could work for your group. Another possibility is to pick the most important topics from the text and assign members of your study group to summarize them during the meetings. These short summaries can initiate discussion of topics that others feel a need to talk about. The person that took the time to prepare the summary will most likely be in the best position to help others with their problems. In this way every member of the group becomes a resource. The group becomes a meeting of tutors!

9. **A tutor can help.** But why fork out the money until you absolutely have to? Students often turn to help from tutors before they have spent good study time on their own. If you have tried the preceding ideas, but still need some personal training, then getting a tutor is a possibility. But don't get a tutor until you have made yourself unwelcome at your instructor's scheduled office hours. Your instructor should be most willing to help you and you should not pay for help until the two of you have determined that it is a good option. Remember that if you communicate with your instructor, then the two of you can work together to make your study efforts pay off. Your instructor might feel comfortable suggesting someone to contact for tutoring. It is usually a violation of school policy for your instructor to offer to tutor you for money. If this happens, talk to the head of the department. A good use of a tutor is to hire someone for your study group (you can split the cost!). Do this right before an exam to help clear up last minute problems with the material.

10. **Avoid some "freshman" mistakes.** The first year of college can be a trying time. It might be your first time away from home. You are now responsible for your own timetable. There is no one but you to decide when things must be done. Scheduling time for adequate study can be a challenge, especially since you are surrounded by new and interesting things: people to meet, places to explore, and activities to take part in. It is a good idea to get involved with your surroundings, but the wise student needs to strike a balance. You might not know how much extracurricular activity you can handle without jeopardizing your grades.

    The best approach is to take it slow. Don't try every new thing that appeals to you. Treat your work like a 40-hour-a-week job. Make sure that the sum of your time in class and your study time is about 40 hours per week. If your schoolwork is getting done easily, then you can add other activities into your schedule. Avoid new things that require a great deal of time. Be critical of your need for a part-time job. Work only if you must do so to pay for your education. Working 30 hours a week to pay your car insurance is not a good strategy if it becomes a threat to good grades (this is a real example from a real student).

11. **It makes a difference how much time you spend studying.** This suggestion is not exactly what it seems. Certainly if you study only a few hours right before an exam, you will not be rewarded with a good grade. No, this recommendation has more to do with the length and timing of your study sessions. For example, make sure that you study your hardest subject *first*. It is only natural to put off to the end of the night your work on the most difficult material. Unfortunately, by the time you get to your difficult studying, you are tired and least likely to learn efficiently.

The most important thing is to study in small bursts. Study each subject *every day*. Read and work through assignments in short sessions (30 minutes to $1\frac{1}{2}$ hours each). A big mistake often made is to "save up" study time for a subject and then try to cram too much learning into one or two marathon sessions before an exam. If you read a little every day and keep ahead of your instructor, you will find it much easier to comprehend the subject and keep up in class. Never delude yourself that you can catch up later when you have no clue about what your instructor is talking about in class. That feeling is a danger sign that should jump-start your studying efforts.

It is wise to write down how you spend *every* hour of the week and then schedule time to study particular subjects. You will be surprised how many hours you waste between classes or events. Use these odd bits of time to your advantage by carrying notes with you to look over and "distill" (more on distilling below). How much study time is reasonable? Often instructors suggest, as a rule of thumb, that for every hour you spend in class per week you should spend two hours studying. However, research on study time reveals that there is little direct correlation between hours of study and performance. If you study little, you'll probably do poorly; but students are so different that it is hard to make a concrete suggestion. You need to start with more hours (not less) and then pare it down as you succeed.

Once you establish your ability to do your class work, socialize, and engage in other activities, you will be more able to judge how much time each new activity will take. In the beginning, however, just stick to your class work and a small amount of socializing until you get your study habits established. Establishing study habits can be hard, especially for someone who did not study much during their high school years. It is important to realize that college level material goes faster and there is less structured help for you in class. If you realize early on that you are on your own to rise to the challenge, you might avoid the biggest "freshman" mistake: wishful thinking. Students with poor study backgrounds often stumble on their first exam and then write it off to bad luck. They predict that things will get better. Of course, they rarely do, unless a serious effort is made to bring study habits up to speed. A bad performance on an exam should sound an alarm to you. See your instructor and restructure your studying to get back on track. The situation will never take care of itself; it is up to *you* to make things happen.

NOTE DISTILLATION

There are two other specific activities that may help you study better. The first involves a new way of reading. The second involves studying with at least one other person. Both have yielded positive results when employed diligently by students. In fact, dramatic improvement has taken place for students that have changed their old habits in part by adopting these new approaches. They might work for you, too.

Remember the admonishment to throw away your highlighting pen? Highlighting material in your text does not cause your brain to absorb it any better. What is worse, most students highlight *too much*, so that when they go back over the text, they are faced with pages of glowing yellow paragraphs. It just doesn't work. But if you are more selective, you will learn more. Instead of reading with a highlighter, keep a separate notebook from the one that contains your class notes. In this notebook write things down as you read. Don't write complete sentences, use phrases instead. Write brief definitions that paraphrase the ones in the text (use your *own* words). Write down formulas and define variables clearly but not elaborately. This will only take you a little longer than reading without taking notes, and it will cause the material to be "seen" more by your brain. You have to process the information more deeply if you are going to write something down.

If a chapter in your text is about twenty pages long, you might have handwritten notes of ten pages. Your goal is to "distill" those pages down to one page. In the end you want one page per chapter to study right before your exam. The best time to distill your notes is during those odd times between classes—during lunch, while you are waiting to talk with your instructor, or whenever you have a few minutes. Carry the notebook with you always since you never know when spare moments will arise.

The purpose of distilling your notes is simple: when you take those ten handwritten pages and distill them to five, you do so by writing down only the things that give you the most difficulty. In other words, you will not write down the things you **know**—the things you have learned. By doing this a few times, you will convert ten pages into one page, which will contain only the things you have the most trouble remembering. It might take you four iterations to pare down your notes, but the process will be surprisingly short. And you will generate a customized study guide. By exam time, you may find that you know your study sheets by heart, which is a wonderful situation.

MOCK EXAMS

Another study method that might work for you requires that you find another student willing to work with you (*Helpful Hint*: Don't choose a boyfriend or girlfriend). It can work for a study group, too. Your task is to write exam questions. Arrange with your partner or group to meet in a neutral, boring place once a week, or however often is appropriate for the coverage in your class. Agree to write a certain number of questions based on the material covered in the textbook. A good number might be five or six questions per chapter. If your class exams will be multiple choice, then make your questions multiple choice. Try to match the way you will be examined for your course.

At the appointed time, exchange practice exams with your partner(s). Give yourselves a reasonable amount of time to do the questions, but no more than you will be allowed for the real exam. After time is called, grade the practice exams. For the problems that are missed, the author must explain the right answers for the benefit of the others in the group. After all the problems have been covered, adjourn to the local café for a bite to eat (if you reward yourself for studying, your Pavlovian instincts will help keep you motivated).

What is the purpose of writing exam questions? Quite simply, if you can write a good question that measures whether someone knows the material, then you must know that material pretty well yourself. This is just a sly way to get you to learn the material well enough so that you can write a legitimate question about it. You also have the added benefit of taking a practice exam to show how well you understand the material. Try to write good questions based on the most important aspects of the material. It might be hard at first, and you will likely get frustrated. Try using the practice examples and homework problems from your text as a guide when you first get started. With practice you will be able to write questions more easily. Remember, the idea behind this method is to get you and your partners more involved in active learning of the material. Writing questions can definitely help.

WEIRD TRICKS

Finally, we will pass along some weird tricks that we know about but don't necessarily know why they work—or if they will work for you.

Weird Trick #1: (This tip sounds absolutely stupid, but some students claim success, and you can't argue with success.) Listen to music the night before an exam (but not the same thing over and over!). During the exam you can "play back" the songs in your head and some of the material might come back to you more easily (we told you it was weird). Studies show that students who study in a certain environment will perform better on tests given under similar circumstances. For example, students that study surrounded by the smell of chocolate will do slightly better on an exam given in a room that has the same smell. This implies that if you study in the room in which your exam will be given, you might have a slight edge during exam time. It's worth a try.

Weird Trick #2: Dress up for the exam. Some students report that if they dress well for an exam, the extra self-confidence that comes from looking good helps them relax and perform better. This isn't like having a lucky pair of socks; it's more like having a lucky evening gown. Give it a try. Some students swear by this one.

Weird Trick #3: Read a novel the night before your final exam. By that time, cramming won't help anyway, so the best thing you can do is relax. Get a good night's sleep. Don't worry with the material the night before the big exam, but look over your study sheets right before you walk into the exam room.

Most of all, change things that don't work. Stay on top of your studies, and when you feel uncomfortable, get help! The way you are studying might not be effective, and you need to modify activities accordingly. Don't try to do it all on your own if you can't seem to do well enough without help. Your instructors really do want you to succeed. Seek help from them as soon as you know that things are not progressing properly.

Chapter 14 Summary

- Your performance in a class depends more on *you* than on your instructor.
- Read your textbook and take notes while you read (throw away your highlighter!).
- Go to class and take notes.
- Do your homework and write down things that are most difficult for you so that your instructors can help you when you go to their office hours.
- Study in small bursts, hardest subject first.
- Use group study effectively (write practice exams or assign summaries to group members).
- Enlist a tutor only after you have exhausted your own arsenal of study aids. Follow the advice of your instructor in this regard.
- Avoid wishful thinking—seek help when you feel unsure of your progress.
- Try distilling your notes to one page per chapter to generate customized study sheets.
- Write exam questions to learn the material in more depth. Trade your questions with a partner to help each other.

APPENDIX

Answers to Practice Exercises

CHAPTER 1

Here are possible approaches. There are other key sequences that will give the right answer. Solutions using a graphing calculator are given in the boxes. Note that in most cases the solution is easier using the graphing calculator.

1. $\Big[$**3.53 – 6.2 = Mem**
   3.53 + 6.2 = ÷ Rec =$\Big]$

   Put denominator into memory.
   Recall denominator from memory.

   ```
   (3.53+6.2)/(3.53
   -6.2)
              -3.644194757
   ```

   Answer: –3.64

2. $\Big[$**4 × 3.5 = – 6 = $^1/_x$**$\Big]$

   Use reciprocal key on denominator.

   ```
   1/(4*3.5-6)
              .125
   ```

   Answer: 0.125

3. $\Big[$**6.2 – 2.6 = log × 2.33 Mem**
   4.12 x^y 3 = – Rec =$\Big]$

   Put second term into memory.
   Use x^y key to do the cube.

Answer: 68.6

4. $\left[4 \times 6 = \text{Mem } 5\ x^2 - \text{Rec} = \sqrt{} + 5 =\right]$

Answer: 6

5. $\left[6.5\ \text{EE } 7 \pm \text{Mem}\right.$ Put denominator into memory.
   $\left.4.7\ \text{EE } 3 \pm + 2.3\ \text{EE } 3 \pm = \div \text{Rec} =\right]$

Answer: $10{,}769.2307 = 1.1 \times 10^4$

(See Chapters 6 and 7 for discussion of scientific notation and significant digits.)

The graphing calculator displays scientific notation if it is chosen under the {MODE} menu (SCI).

6. $\left[3 \times 5 = \text{Mem } 3\ x^2 - \text{Rec} + 6 =\right]$

Answer: 0

On a graphing calculator, type the equation under {Y = }.

Go to {2ⁿᵈ CALC} (above the {TRACE} key), {1:value ENTER 3 ENTER}.

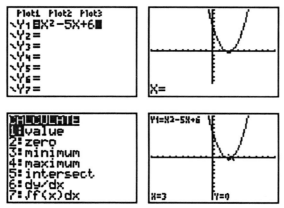

This is an alternate method of evaluating an expression using the table feature.

Go to {2nd **TBLSET**} (next to {Y = }), **Indpnt: Ask, Depend: Auto.**

Go to {2nd **TABLE**}, and type in 3 under the X.

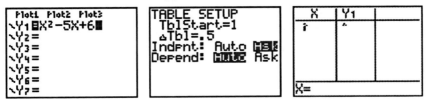

7. [**0.1 − 5 EXP 2 ± = Mem**   Put first part of denominator into memory.

    0.1 + 5 EXP 2 ± = x Rec = Mem   Put denominator into memory.

    2 × 5 EXP 2 ± = x^2   Square the numerator.

    ÷ Rec =]

    Answer: 1.3

    On a graphing calculator, you should follow the same procedure as in exercise 6. The graph looks this way because {**WINDOW**} (the key next to the {Y = }) was adjusted as shown below.

    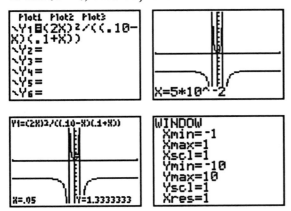

8. Select **radians** mode, then

    [**π ÷ 2 = Mem**   Put $\left(\frac{\pi}{2}\right)$ into memory since it is used twice.

    sin + Rec cos =]   Take **sin** $\left(\frac{\pi}{2}\right)$ after putting $\left(\frac{\pi}{2}\right)$ into memory.

    Answer: 1

    Put your graphing calculator in radian mode first. (Press the {**MODE**} key.)

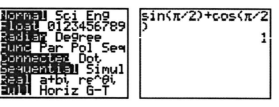

9. Select **degrees** mode, then

   $\left[1.00\,\tan^{-1}\right]$ or $\left[1.00\,\text{INV}\,\tan\right]$

   Answer: 45°

   Put your graphing calculator in degree mode first. (Press the {**MODE**} key.)

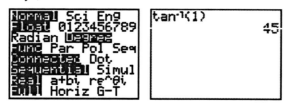

10.

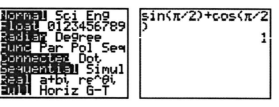

   Go to {**ZOOM**}, 6: **Zstandard**.

   Zstandard changes the window to:

11.   You need to find an appropriate window to see a complete graph.

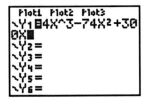 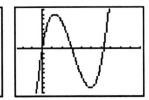

12.   [3 , 4 = **Mem**   Determine the exponent and put it into memory.
       1296 y^x **Rec =**]

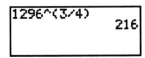

       Answer: 216

13.   [**8.31** × **298** =   Calculate denominator of exponent first.
       ÷ **5000** ⅟ₓ   Divide denominator by numerator and then **invert**.
       $- e^x$]   Change sign on exponent before computing e^x.

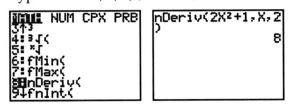

       Answer: 0.133

14.   Go to {**MATH 8:nDeriv(ENTER**}.
       Type function, x, 2, {**ENTER**}.

       Answer: 8

15.   Graph function under Y_1 and Y_2. Go to {**2ⁿᵈ CALC**}, **5:intersect**. Move
       the cursor near the intersection of either function and press {**ENTER**}.
       Use the down arrow key to move the cursor to the other function and
       press {**ENTER**}. Press {**ENTER**} again when the calculator prompts for a
       guess.

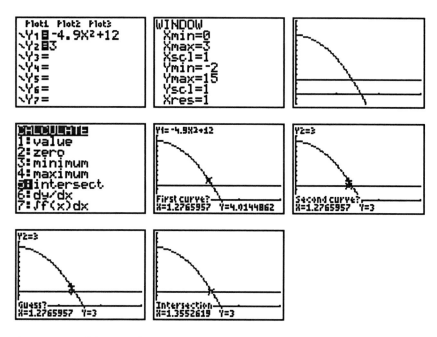

Answer: 1.36 seconds

CHAPTER 2

1. Using a graphing calculator, type in the compound fraction as shown and press {ENTER} to display the decimal. Press {MATH 1: FRAC ENTER} to display the compound fraction.

a. $5\frac{5}{9} = \frac{(9)(5)+5}{9} = \frac{45+5}{9} = \frac{50}{9} = 5.5\overline{5}$

The bar over the 5 in this answer is a **repetend**, which indicates that the 5 repeats indefinitely (your calculator will display a number of digits and round off the last one).

$\left[5 \div 9 = +5 =\right]$ yields $5.5\overline{5}$

```
(9*5+5)/9
             5.555555556
Ans▶Frac
                    50/9
```

b. $2\frac{23}{32} = \frac{(32)(2)+23}{32} = \frac{64+23}{32} = \frac{87}{32} = 2.71875$

```
(32*2+23)/32
           2.71875
Ans▶Frac
             87/32
```

$$\left[23 \div 32 = + 2 =\right]$$

c. $3\frac{5}{11} = \frac{(11)(3)+5}{11} = \frac{33+5}{11} = \frac{38}{11} = 3.\overline{45}$

```
(11*3+5)/11
        3.454545455
Ans▶Frac
             38/11
```

$$\left[5 \div 11 = + 3 =\right]$$

d. $-4\frac{2}{7} = -\left(\frac{(7)(4)+2}{7}\right) = -\left(\frac{(28+2)}{7}\right) = -\frac{30}{7} = -4.28571$

```
-(7*4+2)/7
       -4.285714286
Ans▶Frac
             -30/7
```

$$\left[2 \div 7 = + 4 = \pm\ \right]$$

2. a. Dividing numerator and denominator by 12 yields $\frac{3}{7}$.

   b. Dividing numerator and denominator by 13 yields $\frac{2}{3}$.

   c. Dividing numerator and denominator by 4 yields $\frac{23}{29}$.

   d. Dividing numerator and denominator by 7 yields $\frac{13}{37}$.

3. a. $\frac{13}{8} = \frac{8+5}{8} = 1 + \frac{5}{8} = 1\frac{5}{8} = 1.625$

   $$\left[13 \div 8 =\right]$$

   b. $\frac{305}{137} = \frac{2(137)+31}{137} = 2\frac{31}{137} = 2.22628$

   $$\left[305 \div 137 =\right]$$

c. $-\dfrac{56}{26} = -\dfrac{28}{13} = -\left(\dfrac{2(13)+2}{13}\right) = -2\tfrac{2}{13} = -2.15385$

$\left[56 \div 26 = \pm\right]$

d. $\dfrac{256}{16} = \dfrac{(16)(16)}{16} = 16$

$\left[256 \div 16 =\right]$

4. a. $\tfrac{1}{8} = 0.125$

   $\tfrac{3}{8} = 0.375$

   $\tfrac{5}{8} = 0.625$

   $\tfrac{7}{8} = 0.875$

   b. $\tfrac{1}{3} = 0.3\overline{3}$

   $\tfrac{2}{3} = 0.6\overline{6}$

   $\tfrac{4}{3} = 1.3\overline{3}$ $\quad\left(1\tfrac{1}{3}\right)$

   $\tfrac{5}{3} = 1.6\overline{6}$ $\quad\left(1\tfrac{2}{3}\right)$

5. a. $13.375 = 13 + 0.375 = 13 + \tfrac{3}{8} = 13\tfrac{3}{8} = \tfrac{107}{8}$

```
13.375▶Frac
           107/8
```

b. $7.625 = 7 + 0.625 = 7 + \tfrac{5}{8} = 7\tfrac{5}{8} = \tfrac{61}{8}$

```
7.625▶Frac
            61/8
```

c. $-4.3\overline{3} = -\left(4 + 0.3\overline{3}\right) = -\left(4 + \tfrac{1}{3}\right)$

```
-4.333333333333▶
Frac
           -13/3
```

d. $3.6\overline{6} = 3 + 0.6\overline{6} = 3 + \frac{2}{3} = 3\frac{2}{3}$

```
3.6666666666666▸
Frac
            11/3
```

6. π: Your calculator should have a key that returns π when pressed (it may be a secondary function that requires an extra key to be pressed).
   Answer: $\pi \cong 3.141592654$
   On a TI-83 Plus, the {π} key is over the {$\wedge$} key, so press {2nd $\wedge$ ENTER}.
   e: Determine e as e^1
   Answer: $\left[1\ e^x\right] e \cong 2.718281828$
   On a TI-83 Plus, there is an {e} key over the {+}, so press {2nd + ENTER}. There is also an {e^x} key over {LN}, so press {2nd LN 1 ENTER}.

7. a. -22: integer, rational, real, complex
   b. $\sqrt{36}$: integer, rational, real, complex
   c. -6.2: rational, real, complex
   d. $\sqrt{15}$: irrational, real, complex
   e. $0.\overline{85}$: rational, real, complex
   f. $\frac{3}{2}$: rational, real, complex
   g. $6 + 2i$: complex
   h. $0.70700\ldots$: irrational, real, complex. (It is NOT repeating the same number pattern, so this is not considered a repeating decimal.)

8. Express each in decimal form, then arrange: $\sqrt{4},\ 2\frac{5}{7},\ e,\ \frac{28}{9},\ \pi,\ \sqrt{11}$

9. Express each in decimal form, then arrange:

   $-\frac{7}{2},\ -\sqrt[5]{e^6},\ -\frac{16}{5},\ -\sqrt{10},\ -\pi,\ -3$

10. a. $\dfrac{27}{10},\ \dfrac{54}{20},\ \dfrac{108}{40}$

    $2.7 = 2\dfrac{7}{10} = \dfrac{27}{10} = \dfrac{27 \times 2}{10 \times 2} = \dfrac{54}{20} = \dfrac{54 \times 2}{20 \times 2} = \dfrac{108}{40}$ (Answers may vary!)

    b. $\dfrac{16}{10},\ \dfrac{24}{15},\ \dfrac{32}{20}$

    $\dfrac{8}{5} = \dfrac{8 \times 2}{5 \times 2} = \dfrac{16}{10} = \dfrac{8 \times 3}{5 \times 3} = \dfrac{24}{15} = \dfrac{8 \times 4}{5 \times 4} = \dfrac{32}{20}$

CHAPTER 3 ─────────────

1. a. $20:120 = 2:12 = 1:6$

   $$\frac{20}{20 + 120} = \frac{20}{140} = \frac{1}{7}$$

   b. $13:52 = 1:4$

   $$\frac{13}{13 + 52} = \frac{13}{65} = \frac{1}{5}$$

   c. $2.4:17.2 = 1.2:8.6 = 0.6:4.3 = 6:43$

   $$\frac{2.4}{2.4 + 17.2} = \frac{2.4}{19.6} = \frac{24}{196} = \frac{6}{49}$$

   d. $2:12{,}000 = 1:6{,}000$

   $$\frac{2}{2 + 12{,}000} = \frac{2.4}{12{,}002} = 0.0001666$$

   Note that $\dfrac{2}{12{,}000} = 0.000166\overline{6}$, so $2:12{,}000$ is very nearly equal to

   $\dfrac{2}{12{,}000}$ since $12{,}000$ is so much larger than 2.

2. Solve by cross multiplying and solving for x:
   a. $180x = 180$, $x = 1$
   b. $15x = 18$, divide both sides by 15, $x = 1.2$
3. Move the decimal two places to the right:
   a. 76%   c. 0.46%
   b. 121%   d. 8.4%
4. Move the decimal two places to the left to change to a decimal:
   a. 0.85   c. $32.5\% = 0.325$
   b. 0.082   d. 0.03

Use your graphing calculator to change the decimal to a fraction in lowest terms. Type in the decimal, press {**MATH 1:Frac ENTER**}.

a. $0.85 = \dfrac{85}{100} = \dfrac{85 \div 5}{100 \div 5} = \dfrac{17}{20}$

```
.85►Frac
            17/20
```

b. $0.082 = \dfrac{82}{1000} = \dfrac{82 \div 2}{1000 \div 2} = \dfrac{41}{500}$

```
.082►Frac
            41/500
```

c. $0.325 = \dfrac{325}{1000} = \dfrac{325 \div 25}{1000 \div 25} = \dfrac{13}{40}$

```
.325►Frac
            13/40
```

d. $0.03 = \dfrac{3}{100}$

```
.03►Frac
            3/100
```

5. Divide and move the decimal point two places to the right:

a. 12.5%

```
1/8
            .125
```

b. 40%

```
2/5
            .4
```

c. 57.14%

```
4/7
        .5714285714
```

d. 69.57%

```
16/23
        .6956521739
```

6. a. $\dfrac{4}{121} = \dfrac{x}{100}$, $400 = 121x$, $x = 3.31$, 3.31%

b. $\dfrac{1}{6} = \dfrac{x}{100}$, $6x = 100$, $x = 16.7$, 16.7%

c. $\dfrac{7}{25} = \dfrac{x}{100}$, $25x = 700$, $x = 28$, 28%

d. $\dfrac{5}{9} = \dfrac{x}{100}$, $9x = 500$, $x = 55.56$, 55.56%

7. a. $0.161 \times 180 = 28.98$ c. $0.025 \times 88 = 2.2$

b. $0.75 \times 2552 = 1914$ d. $1.10 \times 6721 = 7393.1$

8. a. $\dfrac{24}{100} = \dfrac{x}{180}$, $100x = 4320$, $x = 43.2$

    b. $\dfrac{85}{100} = \dfrac{x}{12}$, $100x = 1020$, $x = 10.2$

    c. $\dfrac{2}{100} = \dfrac{x}{112}$, $100x = 224$, $x = 2.24$

    d. $\dfrac{125}{100} = \dfrac{x}{426}$, $100x = 53{,}250$, $x = 532.50$

9. Home Fans : Visiting Fans $= 45{,}000 : 5{,}000 = 45 : 5 = 9 : 1$

    $$\% \text{ Visitors} = \frac{\text{visitors}}{\text{visitors} + \text{home fans}} = \frac{5{,}000}{5{,}000 + 45{,}000}$$

    $$= \frac{5{,}000}{50{,}000} = \frac{1}{10} \times 100\% = 0.10 \times 100\% = 10\%$$

    Note that a ratio of 1:9 is 10%, since 1:9 converts to $\dfrac{1}{1+9} = \dfrac{1}{10}$

10. Reptiles = 22%

    Amphibians = 100% – 22% = 78% are amphibians

    Number of amphibians $= (78\%)(74 \text{ animals})$
    $$= 0.78(74) = 57.72 = 58$$

    The number is rounded up since there cannot be fractions of animals.

    $\dfrac{58 \text{ amphibians}}{74 \text{ animals}} = \dfrac{58}{74} = 0.78 = 78\%$

    $\dfrac{16 \text{ reptiles}}{74 \text{ animals}} = \dfrac{16}{74} = 0.22 = \dfrac{22\%}{100\%}$

11. $\frac{4}{5} \times (632 \text{ dentists}) = 505.6 = 506 \text{ dentists}$

    $\frac{4}{5} = 0.8 = 80\%$

    Minority opinion dentists = 100% – 80% = 20%

12. $(50 \text{ weeks})(40 \text{ hours per week}) = (50 \text{ weeks})\left(40\dfrac{\text{hours}}{\text{week}}\right) = 2000 \text{ hours}$

$$\frac{\$47,500}{1 \text{ year}} = \frac{\$47,500}{50 \text{ weeks}} = \frac{\$950}{1 \text{ week}} = \$950 \text{ per week}$$

$$\frac{\$47,500}{1 \text{ year}} = \frac{\$47,500}{2000 \text{ hours}} = \frac{\$23.75}{1 \text{ hour}} = \$23.75 \text{ per hour}$$

13. $\% \text{ impurity} = 100\% - 99.44\% = 0.56\% = \dfrac{0.56}{100}$

$$\frac{0.56}{100}(2000 \text{ lb}) = 11.2 \text{ lb impurities in 1 ton soap}$$

14. $\frac{3}{8} \text{ inch} = 0.375 \text{ inch}$

$$\text{Elongation} = \frac{0.375 \text{ inch}}{120 \text{ inches}} = 0.003125$$

$0.003125 \times 100\% = 0.3125\%$
$0.003125 \times 1000 \text{ ppt} = 3.125 \text{ ppt}$

15. $20 \text{ ppm} = \dfrac{20}{1,000,000}$

$$20 \text{ ppm}\,(10,000,000 \text{ gal}) = \frac{20}{1,000,000}(10,000,000 \text{ gal})$$

$$= \frac{20 \times 10 \times 1,000,000 \text{ gal}}{1,000,000}$$

$$= 20 \times 10 \text{ gal} = 200 \text{ gal}$$

200 gallons of PCB's are dissolved in the reservoir.

CHAPTER 4

1.   a. $100,000 = 10^5$ c. $0.0001 = 10^{-4}$

     b. $512 = 2^9$ d. $625 = 5^4$

2. a. $7^4 = 7 \cdot 7 \cdot 7 \cdot 7 = 49 \cdot 49 = 2401$ **{7 ^ 4 ENTER}**

   b. $(1.5)^3 = (1.5)(1.5)(1.5) = (2.25)(1.5) = 3.375$ **{1.5 ^ 3 ENTER}**

   c. $10^7 = 10 \cdot 10 \cdot 10 \cdot 10 \cdot 10 \cdot 10 \cdot 10 = 10,000,000$ **{10 ^ 7 ENTER}**

   d. $2^{-2} = \dfrac{1}{2^2} = \dfrac{1}{4} = 0.25$ **{2 ^ –2 ENTER}**

   e. $\sqrt[3]{125} = \sqrt[3]{5^3} = 5$ **{125 ^ $\frac{1}{3}$ ENTER}**

   f. $(121)^{\frac{1}{2}} = \sqrt{121} = \sqrt{(11)^2} = 11$ **{121 ^ $\frac{1}{2}$ ENTER}**

3. a. 51.53632   d. 1.162

   b. 0.0356   e. 0.718

   c. 0.5   f. 0.0432

4. a. $2^3 + 2^4 = 8 + 16 = 24$

   b. $10^3 - 10^2 = 1000 - 100 = 900$

   c. $\sqrt{\pi^3 + \pi} = 5.844$ **{2nd $\sqrt{(\pi \wedge 3 + \pi)}$ ENTER}**

   d. $\dfrac{10 - 10^2}{10 + 10^2} = \dfrac{10 - 100}{10 + 100} = \dfrac{-90}{110} = \dfrac{-9}{11} - 0.81\overline{81}$

5. a. $x^{7+4} = x^{11}$ c. $10^{5+\left(\frac{-1}{5}\right)} = 10^{\frac{25-1}{5}} = 10^{\frac{24}{25}}$

   b. $2^{-\pi-1}$ d. $6x^{-4+7}y^6 = 6x^3 y^6$

6. a. $2^{6-3} = 2^3$ c. $x^{7-6}y^{4-2} = xy^2$

   b. $10^{-2-4} = 10^{-6}$ d. $\frac{27}{3}x^{4-1}y^{3-2} = 9x^3 y$

7. a. $\dfrac{1}{10^3}$ c. $x^{-7-3} = x^{-10} = \dfrac{1}{x^{10}}$

   b. $\dfrac{2e^\pi}{7}$ d. $3^{-7-(-2)} = 3^{-5} = \dfrac{1}{3^5}$

8. a. $\sqrt[3]{8} = 2$

    b. $\left(\sqrt{9}\right)^3 = 27$

    c. $\left(\sqrt[3]{64}\right)^{-2} = 4^{-2} = \dfrac{1}{4^2} = \dfrac{1}{16}$

    d. $\left(\sqrt[5]{32}\right)^3 = 2^3 = 8$

9. a. $81^{\frac{1}{2}} = 9$

    b. $3^{\frac{7}{3}}$

    c. $\left(16x^4y^6\right)^{\frac{1}{4}} = 2xy^{\frac{3}{2}}$

    d. $7^{\frac{-3}{6}} = 7^{\frac{-1}{2}} = \dfrac{1}{7^{\frac{1}{2}}} = \dfrac{1}{\sqrt{7}}$

10. a. $3^{2\times4} = 3^8$

     b. $2^3 x^{2\times3} y^3 = 8x^6y^3$

     c. $\dfrac{4^{-2}x^{-2}}{y^{-2}} = \dfrac{y^2}{4^2x^2} = \dfrac{y^2}{16x^2}$

     d. $\left(27^{\frac{1}{3}}\right)^4 = 27^{\frac{4}{3}} = \left(\sqrt[3]{27}\right)^4 = 3^4 = 81$

CHAPTER 5

1. a. Switch x- and y-coordinates:

    $x = \dfrac{1}{3}y + 1$ (original)

    $x - 1 = \dfrac{1}{3}y$

    Multiply both sides by 3:

    $3(x - 1) = y$

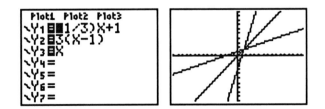

b. Switch x- and y-coordinates:

   $x = 5^y$ (original)

   Use a log to solve for y:

   $y = \log_5 x$

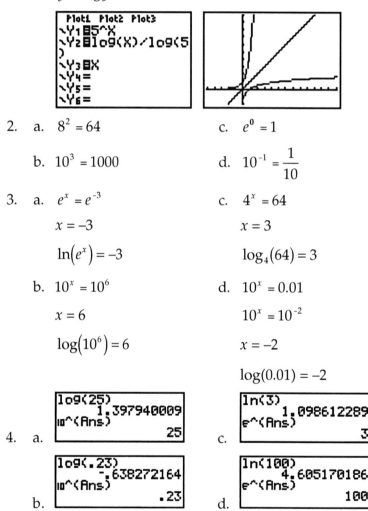

2. a. $8^2 = 64$

    b. $10^3 = 1000$

    c. $e^0 = 1$

    d. $10^{-1} = \dfrac{1}{10}$

3. a. $e^x = e^{-3}$

       $x = -3$

       $\ln(e^x) = -3$

    b. $10^x = 10^6$

       $x = 6$

       $\log(10^6) = 6$

    c. $4^x = 64$

       $x = 3$

       $\log_4(64) = 3$

    d. $10^x = 0.01$

       $10^x = 10^{-2}$

       $x = -2$

       $\log(0.01) = -2$

4. a. ```
        log(25)
               1.397940009
        10^(Ans)
                         25
        ```

    b.  ```
        log(.23)
              -.638272164
        10^(Ans)
                       .23
        ```

    c. ```
        ln(3)
               1.098612289
        e^(Ans)
                          3
        ```

    d.  ```
        ln(100)
               4.605170186
        e^(Ans)
                        100
        ```

5. a. $\log(100 \times 10) = \log(1000)$ d. $\log_4(32 \times 2) = \log_4(64)$

$10^x = 1000$ $4^x = 64$

$x = 3$ $x = 3$

   b. $\log_5\left(\dfrac{75}{3}\right) = \log_5(25)$ e. $\ln\left(\dfrac{e^4}{e^3}\right) = \ln(e)$

$5^x = 25$ $e^x = e$

$x = 2$ $x = 1$

   c. $\log 5^3 + \log 2^3$ f. $\ln\left(\dfrac{e^5}{e^2}\right) = \ln(e^3)$

$\log(125 \times 8) = \log(1000)$ $e^x = e^3$

$10^x = 1000$ $x = 3$

$x = 3$

6. a. $\dfrac{\log(5)}{\log(e)}$ b. $\dfrac{\log(12)}{\log(3)}$

7. a. $\dfrac{\ln(62)}{\ln(10)}$ b. $\dfrac{\ln(17)}{\ln(6)}$

8. $y = \dfrac{\log x}{\log(4)}$

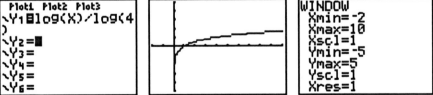

9. $y = \dfrac{\ln x}{\ln (4)}$

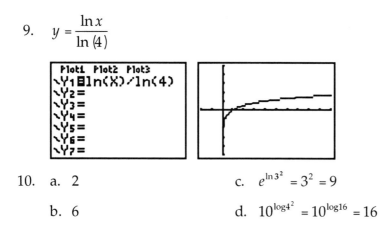

10. a. 2

c. $e^{\ln 3^2} = 3^2 = 9$

b. 6

d. $10^{\log 4^2} = 10^{\log 16} = 16$

CHAPTER 6

These answers are presented without regard to significant figures (see Chapter 7).

1. a. $54{,}000.00 = 5.4 \times 10^4$

c. $-0.00131 = -1.31 \times 10^{-3}$

b. $142.35 = 1.4235 \times 10^2$

d. $0.00000004 = 4 \times 10^{-8}$

2. a. $1.6 \times 10^{-4} = 0.00016$

c. $3.7542 \times 10^3 = 3754.2$

b. $0.2 \times 10^3 = 200$

d. $4.0 \times 10^6 = 4{,}000{,}000$

3. a. $(3.4 \times 10^{-4})(1.7 \times 10^3) = (3.4)(1.7) \times 10^{-4+3} = 5.78 \times 10^{-1} = 0.578$

```
3.4E-4*1.7E3
              .578
```

[3.4 EE 4 ± × 1.7 EE 3 =]

b. $(1.2 \times 10^7)(7.9 \times 10^3) = (1.2)(7.9) \times 10^{7+3} = 9.48 \times 10^{10}$

```
1.2E7*7.9E3
           9.48E10
```

[1.2 EE 7 × 7.9 EE 3 =]

c. $(0.2 \times 10^{-4})(0.4 \times 10^{-4}) = (0.2)(0.4) \times 10^{-4-4} = 0.08 \times 10^{-8} = 8 \times 10^{-10}$

```
.2E-4*.4E-4
          8E-10
```

[0.2 EE 4 ± × 0.4 EE 4 ± =]

d. $(8.4 \times 10^5)(0.5 \times 10^{-9}) = (8.4)(0.5) \times 10^{5-(-9)} = 4.2 \times 10^{-9}$

```
8.4E5*.5E-9
          4.2E-4
```

[8.4 EE 5 × 0.5 EE 9 ± =]

4. a. $\dfrac{3.4 \times 10^{-4}}{1.7 \times 10^3} = \left(\dfrac{3.4}{1.7}\right) \times 10^{-4-3} = 2.0 \times 10^{-7}$

```
3.4E-4/(1.7E3)
          2E-7
```

[3.4 EE 4 ± ÷ 1.7 EE 3 =]

b. $\dfrac{6.5 \times 10^{-4}}{0.5 \times 10^{-6}} = \left(\dfrac{6.5}{0.5}\right) \times 10^{-4-(-6)} = 13 \times 10^2 = 1300$

```
6.5E-4/(.5E-6)
          1300
```

[6.5 EE 4 ± ÷ 0.5 EE 6 ± =]

c. $\dfrac{5.0 \times 10^6}{-2.5 \times 10^{-5}} = \left(\dfrac{5.0}{-2.5}\right) \times 10^{6-(-5)} = -2.0 \times 10^{11}$

```
5E6/(-2.5E-5)
          -2E11
```

[5 EE 6 ÷ 2.5 ± EE 5 ± =]

d. $\dfrac{-4.8 \times 10^{-13}}{1.2 \times 10^5} = \left(\dfrac{-4.8}{1.2}\right) \times 10^{-13-5} = -4.0 \times 10^{-18}$

```
-4.8E-13/(1.2E5)
          -4E-18
```

[4.8 ± EE 13 ± ÷ 1.2 EE 5 =]

5. a. $(2.2 \times 10^{-2}) + (1.4 \times 10^{-2}) = (2.2 + 1.4) \times 10^{-2} = 3.6 \times 10^{-2}$

```
(2.2E-2)+(1.4E-2
)
          .036
        3.6E-2
```

[2.2 EE 2 ± + 1.4 EE 2 ± =]

b. $(5.6 \times 10^4) - (2.1 \times 10^4) = (5.6 - 2.1) \times 10^4 = 3.5 \times 10^4$

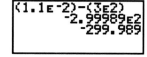

[5.6 EE 4 – 2.1 EE 4 =]

c. $(4.53 \times 10^2) - (0.2 \times 10^3) = (4.53 \times 10^2) - (2. \times 10^2)$

$$= (4.53 - 2.) \times 10^2 = 2.53 \times 10^2$$

[4.53 EE 2 – 0.2 EE 3 =]

d. $(1.1 \times 10^{-2}) - (3.0 \times 10^2) = 0.011 - 300 = -299.989$

[1.1 EE 2 ± – 3.0 EE 2 =]

CHAPTER 7

1. a. 5 significant figures
    b. 4 significant figures
    c. 4 significant figures
    d. 7 significant figures
    e. 3 significant figures
    f. 3 significant figures

2. a. 2.74
    b. 533. or 5.33×10^2
    c. 0.133
    d. 4.22×10^3

3. a. 2.7
    b. 5.3×10^2
    c. 0.13
    d. 4.2×10^2

4. a. 0.0
    b. 3.6×10^{-2}
    c. 0.5×10^3
    d. 37.6
    e. 52.6
    f. 10.979

5. a. 21.4
    b. 5.8×10^1
    c. 0.2×10^4

    d. 7.270×10^{-5}
    e. -6.0×10^6
    f. 7.901×10^3

6. a. 5.7721
    b. 7.57×10^{-4}

    c. 1.695
    d. 1.1

CHAPTER 8

1. a. $r = 11$
    b. $x = -6$

    c. $y = 24$
    d. $a = \dfrac{18}{5}$

2. a. $y = \dfrac{x^2 - 2x}{2} = \dfrac{x^2}{2} - x$

    b. $y = 2x + 1$

    c. $y > 2 - \ln x$

    d. $\ln(x^2) + \ln y = 2$

    $\ln(x^2 y) = 2$

    $e^2 = x^2 y$

    $y = \dfrac{e^2}{x^2}$

    e. $\dfrac{E}{RT} = \ln\left(\dfrac{1}{y}\right)$

    $\dfrac{E}{RT} = \ln(y^{-1})$

    $\dfrac{E}{RT} = -\ln(y)$

$$-\frac{E}{RT} = \ln(y)$$

$$y = e^{-\frac{E}{RT}}$$

  f. $xy + y = 7$, $y(x + 1) = 7$

$$y = \frac{7}{x + 1}$$

3. a. $x < 2$ c. $y \le \dfrac{3}{10}$

     b. $y \le 3$ d. $y > \dfrac{22}{13}$

4. a. $r = \dfrac{d}{t}$ c. $b = \dfrac{2A}{h}$

     b. $w = \dfrac{P - 2l}{2}$ d. $c = \dfrac{1000C}{W}$

5. a. *i.* $(x - 5)(x + 3) = 0$ $x = 5, -3$

       ii. $x = \dfrac{2 \pm \sqrt{4 - 4(1)(-15)}}{2} = \dfrac{2 \pm 8}{2} = 5, -3$

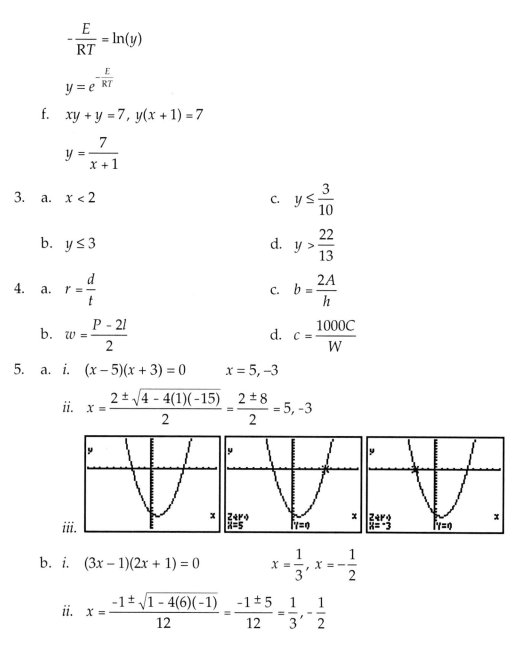

   iii.

     b. *i.* $(3x - 1)(2x + 1) = 0$ $x = \dfrac{1}{3}, \ x = -\dfrac{1}{2}$

       ii. $x = \dfrac{-1 \pm \sqrt{1 - 4(6)(-1)}}{12} = \dfrac{-1 \pm 5}{12} = \dfrac{1}{3}, -\dfrac{1}{2}$

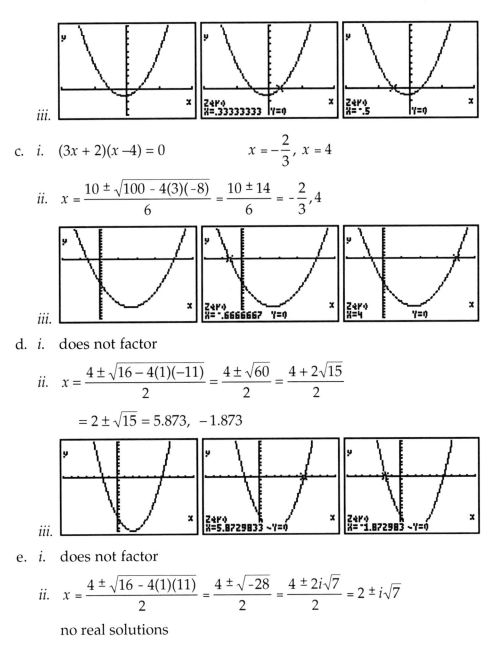

iii.

c. *i.* $(3x + 2)(x - 4) = 0$ $x = -\dfrac{2}{3}$, $x = 4$

   ii. $x = \dfrac{10 \pm \sqrt{100 - 4(3)(-8)}}{6} = \dfrac{10 \pm 14}{6} = -\dfrac{2}{3}, 4$

iii.

d. *i.*   does not factor

   ii. $x = \dfrac{4 \pm \sqrt{16 - 4(1)(-11)}}{2} = \dfrac{4 \pm \sqrt{60}}{2} = \dfrac{4 + 2\sqrt{15}}{2}$

   $= 2 \pm \sqrt{15} = 5.873, \ -1.873$

iii.

e. *i.*   does not factor

   ii. $x = \dfrac{4 \pm \sqrt{16 - 4(1)(11)}}{2} = \dfrac{4 \pm \sqrt{-28}}{2} = \dfrac{4 \pm 2i\sqrt{7}}{2} = 2 \pm i\sqrt{7}$

   no real solutions

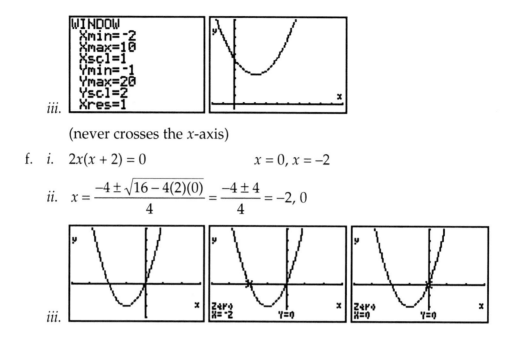

iii.

(never crosses the x-axis)

f. i. $2x(x + 2) = 0$ $x = 0, x = -2$

   ii. $x = \dfrac{-4 \pm \sqrt{16 - 4(2)(0)}}{4} = \dfrac{-4 \pm 4}{4} = -2, 0$

   iii.

CHAPTER 9

1. a. $\$5,000,000 = \$5 \times 10^6 = 5M\$$ (megabucks!)

   b. $2{,}000{,}000{,}000$ bytes $= 2 \times 10^9$ bytes $= 2$ Gbytes or 2 Gb

   c. 0.025 m $= 2.5 \times 10^{-2}$ m $= 2.5$ cm

   d. 4×10^{-9} s $= 4$ ns

   e. 8×10^{-3} W $= 8$ mW (milliwatts)

   f. 10^{-6} m $= 1$ μm (micrometer, or "micron")

   g. $50 \times 10^3 = 50$ kton (kilotons)

   h. 1×10^{12} dactyls $= 1$ Tdactyl (a teradactyl)

2. a. 5.3 Mbar $= 5.3 \times 10^6$ bar, or $5{,}300{,}000$ bar

   b. 0.2 kJ $= 0.2 \times 10^3$ J $= 200$ J

   c. 20 mm $= 20 \times 10^{-3}$ m $= 0.020$ m

d. $1.1 \text{ ng} = 1.1 \times 10^{-9} \text{ g} = 0.0000000011 \text{ g}$

3. a. $(1.0 \text{ yr})\left(365 \, ^{dy}\!/_{yr}\right)\left(24 \, ^{hr}\!/_{dy}\right)\left(60 \, ^{min}\!/_{hr}\right)\left(60 \, ^{s}\!/_{min}\right) = 3.1536 \times 10^{7} \text{ s}$

b. $5.29 \times 10^{-11} \text{ m} = 52.9 \times 10^{-12} \text{ m} = 52.9 \text{ pm}$

c. $\dfrac{(500 \text{ yd})\left(3 \, ^{ft}\!/_{yd}\right)}{5280 \, ^{ft}\!/_{mi}} = 0.284 \text{ mi}$

d. $(0.0025 \text{ mi})\left(5280 \, ^{ft}\!/_{mi}\right)\left(12 \, ^{in}\!/_{ft}\right) = 158.4 \text{ in}$

$\dfrac{(158.4 \text{ in})\left(2.54 \, ^{cm}\!/_{in}\right)}{\left(100 \, ^{cm}\!/_{m}\right)} = 4.0 \text{ m}$

e. $\dfrac{(0.35 \text{ lb})\left(453.6 \, ^{g}\!/_{lb}\right)}{\left(1000 \, ^{g}\!/_{kg}\right)} = 0.16 \text{ kg}$

f. $1.2 \text{ ms} = 1.2 \times 10^{-6} \text{ s}\left(\dfrac{10^{9} \text{ ns}}{1 \text{s}}\right) = 1.2 \times 10^{3} \text{ ns}$

g. $101.3 \text{ MHz} = 101.3 \times 10^{6} \text{ Hz}\left(\dfrac{1 \text{kHz}}{10^{3} \text{Hz}}\right) = 101.3 \times 10^{3} \text{ kHz}$

h. $\dfrac{100 \text{ g}}{453.6 \, ^{g}\!/_{lb}} = 0.220 \text{ lb}$

4. a. $96485 \text{ C} \cdot \text{mol}^{-1}$
    b. $0.08206 \text{ atm} \cdot \text{L} \cdot \text{mol}^{-1} \cdot \text{K}^{-1}$
    c. $2.998 \times 10^{8} \text{ m} \cdot \text{s}^{-1}$
    d. $8.85 \times 10^{-12} \text{ C}^{2} \cdot \text{N}^{-1} \cdot \text{m}^{-2}$

5. a. $1 \, ^{1}\!/_{s}$
    b. $9.81 \, ^{m}\!/_{s^{2}}$
    c. $8.314 \, \dfrac{\text{J}}{\text{mol} \cdot \text{K}}$
    d. $1 \, \dfrac{\text{kg} \cdot \text{m}^{2}}{\text{s}^{2}}$

CHAPTER 10 ─────────────────

1. a. This is not a straight line, there is a curved part that flattens out:

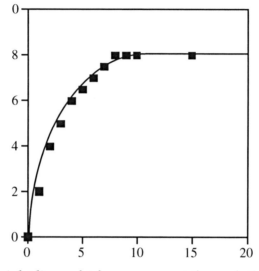

b. This is a straight line, which you can put through the points by eye:

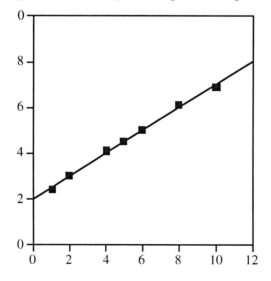

2. a. This line is drawn easily using the equation:

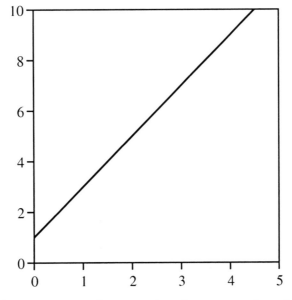

b. You need to rearrange the equation first to $y = \frac{1}{2}x - 1$:

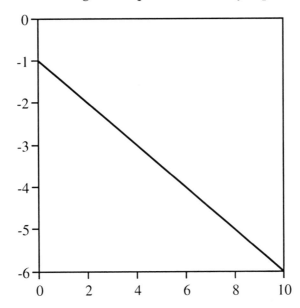

3.   1(b) The intercept is near 2. The line follows the equation $y = \frac{1}{2}x + 2$.

2(b) The intercept is -1 [see preceding answer to 2(b)].

4. a. 3.5   d. 1.46

    b. 9   e. 0

    c. –7   f. –12

5. A table for Least Squares analysis looks something like this:

x	y	x^2	xy
1	2.4	1	2.4
2	3.0	4	6.0
4	4.1	16	16.4
5	4.5	25	22.5
6	5.0	36	30.0
8	6.1	64	48.8
10	6.9	100	69.0
Sum 36	32.0	246	195.1

$$m = \frac{7(195.1) - (36)(32)}{7(246) - (36)^2} = 0.502$$

$$b = \frac{(246)(32) - (36)(195.1)}{7(246) - (36)^2} = 1.992$$

6.

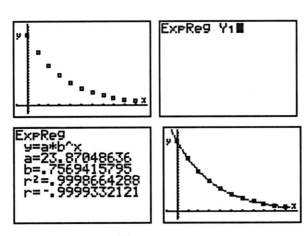

$$y = 23.87(0.7569)^x$$

7.   Use the slope-intercept form of a line.

   a. $y = 5x + 4$

   b. $y = \dfrac{2}{3}x - 2$

8.   Use the point-slope form of a line.

   a. $y - 5 = \dfrac{1}{3}(x - 1)$ b. $m = \dfrac{5 - 3}{-2 - (-1)} = -2$

   $y = \dfrac{1}{3}x + \dfrac{14}{3}$ $y - 3 = -2(x - (-1))$

                           $y = -2x - 1$

9. The two graphs look virtually identical.

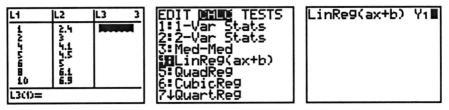

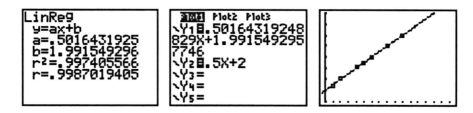

CHAPTER 11

1. a. $\pi = 180°$

    b. $\frac{\pi}{4} = 45°$

    c. $\frac{-3\pi}{2} = -270°$

    d. $2.43\left(\dfrac{180°}{\pi \text{ radians}}\right) = 139°$

2. a. $135° = \frac{3\pi}{4}$

    b. $-270° = \frac{-3\pi}{2}$

    c. $30°\left(\dfrac{\pi \text{ radians}}{180°}\right) = \frac{\pi}{6}$ radians

    d. $257.3°\left(\dfrac{\pi \text{ radians}}{180°}\right) = 4.491$ radians

3. a. 0

    b. 0

    c. $\sin(45°) = \sqrt{\dfrac{1}{2}} = \dfrac{\sqrt{2}}{2}$

    d. $\dfrac{\sin(60°)}{\cos(60°)} = \dfrac{\frac{\sqrt{3}}{2}}{\frac{1}{2}} = \sqrt{3}$

4. a. $\cos(23°) = \dfrac{2.5}{H}$ $H = 2.716$

    b. $\cos\theta = \dfrac{A}{H} = \dfrac{3.0}{4.0} = 0.75$

       $\theta = \cos^{-1}(0.75) = 41.4°$

    c. $(2.2)^2 + (4.4)^2 = c^2$, c =4.92

    d. $\sin(62°) = \dfrac{O}{5.0}$
       Opposite = 5.0×sin(62°) = 4.415
       $\cos(62°) = \dfrac{A}{5.0}$
       Adjacent = 5.0¥×os(62°) = 2.347
       Angles = 62°, 90°, and 90° − 62° = 28°

5. a. $A_x = -4$, $A_y = 3$, $|A| = \sqrt{(-4)^2 + (3)^2} = \sqrt{25} = 5$

    $$\theta = \tan^{-1}\left(\dfrac{|A_y|}{|A_x|}\right) = \tan^{-1}\left(\dfrac{3}{4}\right) = 36.9°$$

b. $\vec{A} + \vec{B} = (-2,2) + (4,1) = (-2+4, 2+1) = (2,3)$

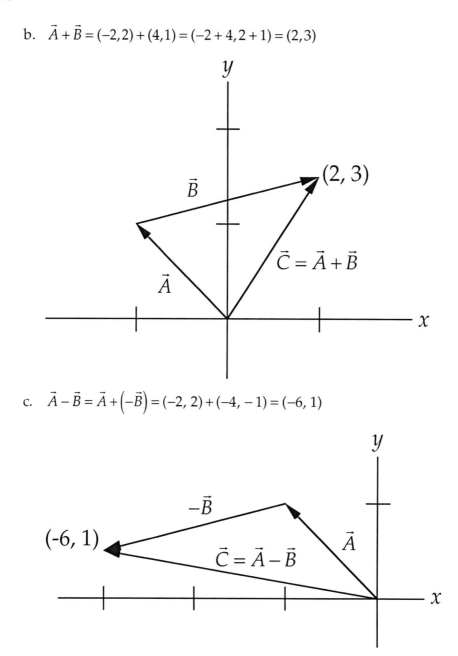

c. $\vec{A} - \vec{B} = \vec{A} + \left(-\vec{B}\right) = (-2, 2) + (-4, -1) = (-6, 1)$

6. a. $P = 4 \times \text{side} = 4(4.2 \text{ cm}) = 16.8 \text{ cm}$

   $A = (\text{side})^2 = (4.2 \text{ cm})^2 = 17.64 \text{ cm}^2 = 1.8 \times 10^1 \text{ cm}^2$

   b. $A = \pi r^2 = \pi \left(\dfrac{d}{2}\right)^2 = \pi \left(\dfrac{6.0 \text{ in}}{2}\right)^2 = \pi(3.0 \text{ in})^2 = 28.27 \text{ in}^2 = 2.8 \times 10^1 \text{ in}^2$

   c. $V = l \times w \times d = 2 \times 4 \times 5 = 8 \times 5 = 40 \text{ units}^2$

   d. $V = h \times \pi r^2 = (10.0)(\pi)(2.0)^2 = 1.3 \times 10^2 \text{ units}^2$

   e. $V = \frac{4}{3}\pi r^3 = \frac{4}{3}(\pi)(1 \times 10^{-6} \text{ m})^3 = 4.2 \times 10^{-18} \text{ m}^3$

   f. $A = \frac{1}{2}(b \times h) = \frac{1}{2}(10.0 \text{ in})(2.00 \text{ in}) = 10.0 \text{ in}^2$

7. a. short leg $= (\text{hypotenuse})/2 = 3 \text{ in}$;

   long leg $= \sqrt{3} \times (\text{short leg}) = 3\sqrt{3} \text{ in}$

   b. short leg $= (\text{long leg})/\sqrt{3} = 6\sqrt{3}/\sqrt{3} = 6 \text{ cm}$;

   hypotenuse $= 2 \times (\text{short leg}) = 12 \text{ cm}$

   c. legs $= (\text{hypotenuse})/\sqrt{2} = 8/\sqrt{2} = 4\sqrt{2} \text{ ft}$

   d. hypotenuse $= (\text{leg}) \times \sqrt{2} = (2\sqrt{2}) \times \sqrt{2} = 4 \text{ m}$

8. $\sin\theta = \dfrac{3}{5}$ $\qquad\qquad \csc\theta = \dfrac{5}{3}$

   $\cos\theta = \dfrac{-4}{5}$ $\qquad\qquad \sec\theta = \dfrac{-5}{4}$

   $\tan\theta = \dfrac{-3}{4}$ $\qquad\qquad \cot\theta = \dfrac{-4}{3}$

9. a. $\theta = \dfrac{\pi}{6}, \dfrac{5\pi}{6}$ $\qquad$ c. $\theta = \pi$

   b. $\theta = \dfrac{\pi}{4}, \dfrac{5\pi}{4}$ $\qquad$ d. $\theta = \dfrac{\pi}{4}, \dfrac{7\pi}{4}$

10. a. $\sin\theta = \dfrac{4}{\sqrt{65}}$ $\qquad$ $\csc\theta = \dfrac{\sqrt{65}}{4}$

$\qquad$ $\cos\theta = \dfrac{7}{\sqrt{65}}$ $\qquad$ $\sec\theta = \dfrac{\sqrt{65}}{7}$

$\qquad$ $\tan\theta = \dfrac{4}{7}$ $\qquad$ $\cot\theta = \dfrac{7}{4}$

b. If θ is in the third quadrant:

$\qquad$ $\sin\theta = \dfrac{-4}{\sqrt{65}}$ $\qquad$ $\csc\theta = \dfrac{-\sqrt{65}}{4}$

$\qquad$ $\cos\theta = \dfrac{-7}{\sqrt{65}}$ $\qquad$ $\sec\theta = \dfrac{-\sqrt{65}}{7}$

$\qquad$ $\tan\theta = \dfrac{4}{7}$ $\qquad$ $\cot\theta = \dfrac{7}{4}$

CHAPTER 12

1. $m = 22.0$ g

$V = 14.3 \text{ cm}^3 - 10.0 \text{ cm}^3 = 4.3 \text{ cm}^3$

$\text{Density} = \dfrac{\text{Mass}}{\text{Volume}} = \dfrac{22.0 \text{ g}}{4.3 \text{ cm}^3} = 5.1 \text{ g/cm}^3$

This is not gold. Pyrite ("fool's gold") has a density near 5 g/cm^3.

2. $c = 2.998 \times 10^8 \text{ m/s} = \dfrac{\text{distance}}{\text{time}}$

time of pulse = 5 ns

distance = (time)(c) = $(5 \times 10^{-9} \text{ s})(2.998 \times 10^8 \text{ m/s}) = 1.5$ m (about 5 ft)

For a shorter pulse, time = 2 ps = 2×10^{-12} s

distance = $(2 \times 10^{-12} \text{ s})(2.998 \times 10^8 \text{ m/s}) = 6 \times 10^{-4}$ m = 0.006 cm = 0.02 in

3. $R_{marble} = 5.0$ mm

$D_{marble} = 2(R) = 10$ mm $= 1.0$ cm

$D_{tire} = 10.0$ ft

Circumference of tire $= \pi D = \pi(10.0$ ft$) = 31.4$ ft

The number of marbles will be equal to $\dfrac{\text{Circumference}_{tire}}{\text{Diameter}_{marble}}$:

$C_{tire} = (31.4$ ft$)(12$ in$/$ft$)(2.54$ cm$/$in$) = 957$ cm

number of marbles $= \dfrac{C_{tire}}{D_{marble}} = \dfrac{957 \text{ cm}}{1.0 \text{ cm}} = 957$ marbles

4. $v_{sound} = 1140$ ft$/$s

distance $= (5.0$ mi$)(5280$ ft$/$mi$) = 26{,}400$ ft

$\text{Velocity} = \dfrac{\text{Distance}}{\text{Time}}$

$\text{Time} = \dfrac{\text{Distance}}{\text{Velocity}} = \dfrac{26400 \text{ ft}}{1140 \text{ ft}/\text{s}} = 23.2$ s

It is relatively easy to estimate seconds between a lightning strike and the thunder clap (count 1-Mississippi-2-Mississippi, etc.). An equation which gives a relation between each second counted and the distance would be handy:

Distance (in feet) = time (in seconds) $\times 1140$ ft$/$s

Distance (in miles) = time (in seconds) $\times 0.22$ mi$/$s

Since 0.22 is close to one-fifth, a good estimate would be

$\text{Distance (in miles)} = \dfrac{\text{Time (in seconds)}}{5}$

5. $h = 6.63 \times 10^{-34}$ J $\cdot$ s

$m = 90$ g $= 0.090$ kg   (need kg for equation)

$v = \dfrac{(90 \text{ mi}/\text{hr})(5280 \text{ ft}/\text{mi})(12 \text{ in}/\text{ft})(2.54 \text{ cm}/\text{in})}{(60 \text{ min}/\text{hr})(60 \text{ s}/\text{min})(100 \text{ cm}/\text{m})} = 40 \text{ m}/\text{s}$

$$\lambda = \frac{h}{mv} = \frac{6.63 \times 10^{-34} \text{ J} \cdot \text{s}}{(0.090 \text{ kg})(40 \text{ m/s})} = 1.8 \times 10^{-34} \frac{\text{kg} \cdot \text{m}^2 \cdot \text{s}^{-2} \cdot \text{s}}{\text{kg} \cdot \text{m} \cdot \text{s}^{-1}}$$

1.8×10^{-34} m, an extremely short wavelength

6. Area(wall) = 9 ft × 9 ft = 81 ft^2

   Area(roll) = 1.5 ft × 20 ft = 30 ft^2

   $$\text{number of rolls} = \frac{\text{Area(wall)}}{\text{Area(roll)}} = \frac{81 \text{ ft}^2}{30 \text{ ft}^2} = 2.7$$

   You must buy three rolls.

7. Area(rink) = 200 ft × 50 ft = 10,000 ft^2

   Density(ice) = 0.92 g/cm^3 = mass/volume

   Need the volume of ice, determined from the area and thickness:

   $$\text{Volume} = \text{Area} \times \text{Thickness}$$

   Since the density is given in g/cm^3, we need to get volume in cm^3.

   Get it in in^3 first, after converting the area to in^2:

   Area = (10000 ft^2)(12 in/ft)^2 = 1.44 × 10^6 in^2

   Note that the conversion factors were squared.

   V = (1.44 × 10^6 in^2)(1.0 in) = (1.44 × 10^6 in^3)(2.54 cm/in)^3 = 2.36 × 10^7 cm^3

   Mass = (0.92 g/cm^3)(2.36 × 10^7 cm^3) = 2.2 × 10^7 g

   $$= \frac{2.2 \times 10^7 \text{ g}}{453.6 \text{ g/lb}} = 4.8 \times 10^4 \text{ lb}$$

   The ice weighs 48,000 pounds, or 24 tons.

8. Flour: $\dfrac{300 \text{ g}}{28 \text{ }^g\!/_{oz}} = 10.7 \text{ oz}$

   Butter: $\dfrac{180 \text{ g}}{28 \text{ }^g\!/_{oz}} = 6.4 \text{ oz}$

   Sugar: $\dfrac{70 \text{ g}}{28 \text{ }^g\!/_{oz}} = 2.5 \text{ oz}$

   Temperature:

   $$T_{lower} = \left(\frac{9 \text{ °F}}{5 \text{ °C}}\right)(170 \text{ °C}) + 32 \text{ °F} = 338 \text{ °F}$$

   $$T_{upper} = \left(\frac{9 \text{ °F}}{5 \text{ °C}}\right)(190 \text{ °C}) + 32 \text{ °F} = 374 \text{ °F}$$

   Set the oven at around 350 °F.

9. 14.7 psi = 14.7 lb air over 1 in^2 of Earth

   For 1 mi^2 base, $(5280 \text{ }^{ft}\!/_{mi})^2 (12 \text{ }^{in}\!/_{ft})^2 = 4.01 \times 10^9 \text{ in}^2$

   Set up proportion:

   $$\frac{14.7 \text{ lb}}{1 \text{ in}^2} = \frac{x \text{ lb}}{4.01 \times 10^9 \text{ in}^2}$$

   $x = (4.01 \times 10^9 \text{ in}^2)(14.7 \text{ }^{lb}\!/_{in^2}) = 5.9 \times 10^{10} \text{ lb of air}$

   The maximum amount of argon is:

   100% − 75.5% N_2 − 23.1% O_2 = 1.4% argon

   The maximum weight of argon is thus

   $1.4\%(5.9 \times 10^{10} \text{ lb}) = 0.014(5.9 \times 10^{10} \text{ lb})$

   $$= 8.3 \times 10^8 \text{ lb} = 410{,}000 \text{ tons argon}$$

10. $(10 \text{ furlongs})(10 \text{ }\tfrac{chains}{furlong})(100 \text{ }\tfrac{links}{chain})(7.92 \text{ }\tfrac{in}{link}) = 79{,}200 \text{ in}$

    $\dfrac{79{,}200 \text{ in}}{12 \text{ }^{in}\!/_{ft}} = 6600 \text{ ft} = 1\tfrac{1}{4} \text{ mile}$

CHAPTER 13 ────────────────

1. Volume = 5 ft × 5 ft × 10 ft = 250 ft^3

    $$\text{Density} = \frac{\text{Mass}}{\text{Volume}}$$

    Mass = (Volume)(Density)

    $= (250 \text{ ft}^3)(12 \text{ in/ft})^3(2.54 \text{ cm/in})^3(2.5 \text{ g/cm}^3)$

    $\cong (250)(2000)(2.5)^3(2.5) \text{ g} \cong (250)(2000)(2.5)^2(2.5)^2 \text{ g}$

    $\cong (500{,}000)(6)(6) \text{ g} \cong (1{,}000{,}000)(3)(6) \text{ g} \cong 18{,}000{,}000 \text{ g}$

    $$\frac{18{,}000{,}000 \text{ g}}{454 \text{ }^{g}\!/\!_{lb}} \cong \frac{(900)(20{,}000)}{450} \cong 40{,}000 \text{ lb}$$

    $$\frac{40{,}000 \text{ lb}}{2{,}000 \text{ }^{lb}\!/\!_{ton}} = 20 \text{ tons of bauxite}$$

2. Mass: $\dfrac{(158 \text{ grain})(454 \text{ }^{g}\!/\!_{lb})}{7000 \text{ }^{grain}\!/\!_{lb}} \cong \dfrac{(160)(450)}{(7000)} \text{ g}$

    $\cong \dfrac{(16)(45)}{70} \cong \dfrac{(16)(9)}{14} \cong 9 \text{ g bullet } = 0.009 \text{ kg}$

    Velocity: $\left(2500 \text{ }^{ft}\!/\!_{s}\right)\left(12 \text{ }^{in}\!/\!_{ft}\right)\left(2.54 \text{ }^{cm}\!/\!_{in}\right) \cong (2500)(24+6) \text{ }^{cm}\!/\!_{s}$

    $\cong (2500)(30) \text{ }^{cm}\!/\!_{s} \cong 75{,}000 \text{ }^{cm}\!/\!_{s} \cong 750 \text{ }^{m}\!/\!_{s} \text{ bullet velocity}$

    Energy: $E = \frac{1}{2}mv^2 \cong (0.5)(0.009 \text{ kg})\left(750 \text{ }^{m}\!/\!_{s}\right)^2$

    $\cong (0.5)(0.009 \text{ kg})(75)(75)(10)(10) \text{ }^{m}\!/\!_{s^2}$

    $\cong (0.5)(0.009)(75)(75)(100) \text{ }\frac{kg \cdot m^2}{s^2}$

    $\cong (0.5)(0.9)(75)(75) \text{ J} \cong (0.5)(75)(75) \text{ J}$

    $\cong (0.5)(3)(25)(3)(25) \text{ J} \cong (0.5)(9)(625) \text{ J}$

    $\cong (0.5)(6250) \text{ J} \cong 3120 \text{ J} \cong 3 \text{ kJ} \quad \text{(a bit high)}$

3. $\text{pH} = -\log(0.004) = -\log\left(4 \times 10^{-3}\right)$

since $\log(5) = 0.7$, estimate that $\log(4) \cong 0.6$, so

$\text{pH} = -[\log(4) + \log(10^{-3})] \cong -[0.6 - 3] \cong -(-2.4) \cong 2.4$

4. $1 \text{ ft}^3 = (12 \text{ in})^3 = (144)(12) \text{ in}^3$

20 beans in 1 in^2 means the area $L^2 = 20$ beans, so $L = \sqrt{20} \cong 4.5$ beans

In 1 in^3 there should be $(4.5)(20) \cong 90$ beans per cubic inch

$\text{Total beans} \cong \left[(144)(12)\text{in}^3\right] \times \left[\frac{90 \text{ beans}}{1 \text{ in}^3}\right] \cong (144)(12)(90)$

$$\cong (144)(1000) \cong 144{,}000 \text{ beans} \quad \text{(that's a lot of beans)}$$

5. a. Round trip

$$\frac{1078 \text{ mi} \times 1.55 \text{ \$/gal}}{28 \text{ mi/gal}} \cong \frac{(1100)(1.6)}{28} \cong \frac{1100 + 660}{30} \cong \frac{1760}{30} \cong \frac{176}{3} \cong \$60$$

b. $\dfrac{1078 \text{ mi}}{65 \text{ mi/hr}} \cong \dfrac{1100 \text{ mi}}{65 \text{ mi/hr}} \cong \dfrac{220}{13} \cong \dfrac{225}{12.5} \cong 17 \text{ hr}$

$17 \text{ hr} + 4(20 \text{ min}) = 18 \text{ hr } 20 \text{ min}$, so a round trip $= 36 \text{ hr } 40 \text{ min}$

6. $\dfrac{2 \times 25 \text{ m} \times 100 \text{ cm/m}}{2.54 \text{ cm/in}} \cong \dfrac{2 \times 2500 \text{ in}}{2.5} \cong 2000 \text{ in per lap}$

$(5280 \text{ ft/mi})(12 \text{ in/ft}) = (5280)(12) \text{ in/mi}$

$\text{Laps} \cong \dfrac{(5280)(12) \text{ in/mi}}{2000 \text{ in/lap}} \cong (5.3)(6) \text{ lap/mi} \cong 30 + \frac{1}{3}(6) \cong 32 \text{ laps in a mile}$

INDEX

188

Printed in the United States
116671LV00004B/1-150/A

9 780471 270546